OUT IN FRONT

Effective Supervision in the Workplace

OUT IN FRONT

Effective Supervision in the Workplace

Leslie E. Dennis
Meredith L. Onion

National Safety Council

Technical Advisor: Carlton D. Piepho
Project Editor: Patricia M. Laing
Cover design: Bob Sunyog
Composition: Media Graphics Corporation

©1990 by the National Safety Council
All rights reserved.
Printed in the United States of America.
94 93 92 91 90   5 4 3 2 1

Library of Congress Cataloging in Publication Data
National Safety Council
Out in Front: Effective Supervision in the Workplace
International Standard Book Number: 0-87912-144-0
Library of Congress Catalog Card Number: 89-060471
5M390   Product Number 15131

Dedicated to Barbara and Dave
for all your support.

Les and Meredith

Contents

9   Interviewing, or Talking It Over   **125**

10   Performance-Based Supervision: Supervisors and Performance Appraisals   **149**

11   Discipline in the Workplace   **166**

Preface

OUT IN FRONT: EFFECTIVE SUPERVISION IN THE WORKPLACE was written for Supervisors and designed to be used as (1) a handbook for the classroom or self-study and (2) a resource for problem-solving in the workplace. The authors discuss exactly what Supervisors must do to be effective in their work—the three key aspects of superviosry responsibilities:

- Supervising those who work for them effectively, efficiently and productively (human resources)

- Supervising the flow of material, goods and services effectively, efficiently and productively (physical resources)

- Carrying out both of these responsibilities safely and effectively and assuring that others do the same

The focus is primarily on Superisors and workers—on the human resources of an organization. The human resources are the most critical component of supervisory responsibilities, and working safely and effectively as a team is the most challenging and rewarding task a Supervisor faces in today's workplace.

The philosophical basis of *OUT IN FRONT* is pragmatic and behavioral. It is based on two key concepts:

- Supervisors must be concerned with and master those things they need to know and do, as opposed to those things they may be concerned with when or if they are promoted to a managerial position.

- When dealing with employees, Supervisors must always focus on the acceptable or unacceptable aspects of employee work behavior (performance) rather than on trying to be an employee counselor or psychologist.

How to Use This Book

The fourteen chapters of *OUT IN FRONT: EFFECTIVE SUPERVISION IN THE WORKPLACE* are organized to provide a logical progression of the duties and concerns of Supervisors and to build on the skills involved. Each chapter stands alone and can be taught as an individual unit. The chapters also build on each other, and contain chapter cross-references to additional material on major subjects. Each chapter contains examples and/or a case study that illustrates major issues or skills covered in the chapter. The case studies are drawn from real supervisory situations and provide questions for discussion.

About the Authors

OUT IN FRONT: EFFECTIVE SUPERVISION IN THE WORKPLACE was written for the National Safety Council by Mr. Leslie E. Dennis and Ms. Meredith L. Onion. Dennis has years of experience working with and training Supervisors in many U.S. and international industries, including railroads, airlines, insurance, banking, financial exchanges, heavy and light manufacturing, hospitals, video, computer, advertising, law, trade unions, bakeries, trade associations, and bottlers. Dennis is a Lecturer in Industrial Relations and Director of External Programs at the Institute of Industrial Relations, Loyola University Chicago. Onion is a specialist in Employee Relations and has conducted training in management development. She is the Director, Employee Relations and Records Management, American Medical Association.

Chapter 1

Introduction: What This Book Is About

OVERVIEW

Supervisors are made, not born. How they are developed, the process of creating a good Supervisor and a leader of the workplace who possesses all necessary skills, is what this book is about. This is a book for Supervisors about their work, about how to effectively supervise. When we use the term Supervisor, we use it to cover those people who have direct-line responsibility for the management of other employees. Sometimes such people are called foremen, sometimes even managers. We take the liberty of using Supervisor to cover it all. It appears to be the most comprehensive term.

The National Labor Relations Act defines a Supervisor as:

Any individual having authority, in the interest of the employer, to hire, transfer, suspend, lay off, recall, promote, discharge, assign, reward, or discipline other employees, or responsibility to direct them, or to adjust their grievances, or effectively to recommend such action, if in connection with the foregoing the exercise of such authority is not of a merely routing or clerical nature, but requires the use of independent judgment.

That's quite a mouthful! Of course, the National Labor Relations Board uses that definition primarily for the purpose of determining if a person is excluded from a union bargaining unit because of supervisory duties. Most Supervisors do not exercise all of those responsibilities, but they do exercise some of them.

What is the difference between the definition of a Supervisor and a manager? The American Management Association says that management is the matter of:

Guiding human and physical resources into dynamic organization units that attain their objectives to the satisfaction of those served and with a high degree of morale and sense of attainment on the part of those rendering the service.

If that strikes you as a lot of gobbledygook, you're not alone. Sorting it all out, Supervisors are dealing with the day-in, day-out problems of handling people and goods on the front line of the workforce. Managers, on the other hand, are at least two steps removed from the front line, since they are, first, managing Supervisors. Sometimes, of course, they're also managing administrative assistants, assistant managers, other managers, assistant vice presidents, vice presidents or even senior vice presidents. When they manage people above that level, we usually call them presidents or chief executive officers.

Our reading of the literature on supervision shows that a great deal of the material covers subjects that have little to do with the day-to-day work of Supervisors. This includes subjects ranging from profit margins to strategic planning to organizational development. Those are things that managers deal with (or should). A lot of books on supervision are simply rewritten books on management. They have completely failed to recognize that the work of the Supervisor is different from the work of the manager, and that Supervisors are very important to the organization in and of themselves.

There are some responsibilities that both managers and Supervisors share, such as dealing with people, handling people problems, meeting goals and keeping things moving. There are other things, such as long-term strategic planning, that are clearly the province of management. Much of what a manager does relates to areas of work or business—finance, marketing, planning, sales, supplies, product improvement, productivity—rather than directly impacting on the production of goods or services.

The Supervisor, on the other hand, is caught up almost exclusively with the problems of keeping people directly engaged in the production of goods and services. Most of the time, the Supervisor is dealing with people and people problems. Robots need technicians to keep them working. People need Supervisors.

With these distinctions in mind, we have tried to keep things on one focus: the human resource side of the job of the Supervisor. This book will cover all aspects of that topic, including:

Chapter 2. The Front-Line Roles of the Supervisor
Chapter 3. Leadership: The Evolving Job of the Supervisor
Chapter 4. Communication: A Two-Way Street
Chapter 5. Talking to the Troops, or Talking with the Team
Chapter 6. Supervising a Diverse Workforce
Chapter 7. Supervisors and the Law: Employment and Workplace Law
Chapter 8. Supervision in a Union Environment
Chapter 9. Interviewing, or Talking It Over
Chapter 10. Performance-Based Supervision: Supervisors and Performance Appraisals
Chapter 11. Discipline in the Workplace
Chapter 12. Planning, Decision-Making and Overcoming Resistance to Change
Chapter 13. Developing Employee Skills and Careers
Chapter 14. Developing Your Career

Let's turn now to some of the aspects of the Supervisor's job.

THE HUMAN RESOURCE JOB OF THE SUPERVISOR

The most important job of the Supervisor in today's workplace is dealing with people. We call this subject human resource management. Since people are the main resource the Supervisor deals with, it's no wonder the human resource management aspect is so important. While the title might sound like jargon, it does make us focus on people as organization resources, and that is a very helpful way to look at them. It does express the fact that today's Supervisor is expected to be a lot more than the front-line boss.

Dealing with Resources

The Supervisor is a manager of resources. If you think about it, no organization or corporation can function without resources. A company must have a building or office, equipment such as desks, machines, computers, labs, lathes, vehicles, whatever. These are the material resources. But, to make all this work, there must also be people: the human resources. Even in a state-of-the-art robotics factory there are people involved.

Looking at it that way, machines, buildings, vehicles, computers, etc. are one part of the resources used by all organizations, and people are the other part. Which do you think is more important? Can the assembly line produce things without people? Of course not. Can people produce things without an assembly line? Yes, though perhaps not as well or as many as are needed. While this is a very simple way of looking at things, it does make clear the central importance of our human resources.

The Supervisor, standing out in front of all the other managers in the company, often feeling very unmanagerial, is really the key contact person in dealing with the human resources of the organization. If you're going to get organized and safe production out of a workplace, no organization or corporation can function without resources, including material and human resources. The Supervisor is the front-line manager of human resources!

Production, Productivity and the Supervisor

Another important human resource management function of the Supervisor deals with safe production and productivity. While production and productivity are different concepts, they cover some of the same ground, and they always involve people management. At its simplest level,

Production is the amount of product or services that can be produced in a company.

Of course, that definition doesn't account for how long it took to produce those goods or services or the quality of the product, so you soon begin changing the definition to something like:

Production is the amount of quality product or services that can be safely produced in a company over a period of an hour, day, week, month or year.

Now let's start thinking about how to increase production. We begin by looking at the three major elements necessary for that production: materials, machines and people. These are the basic ingredients of productivity, that is, improving the volume of production in a given period of time through the use of better materials, more efficient/faster machines, and people working more efficiently and safely. Again, we must remember that better materials and faster machines still won't increase productivity without people running them and Supervisors managing those people, materials and machines. Since the Supervisor usually isn't the one who orders the materials or machines (or the computers, telephones, lab equipment, trucks, etc.) the Supervisor's primary role in production and productivity becomes pretty clear. It is largely a matter of:

Managing the people who produce the goods or services.

In fact, the Supervisor is probably the most important element in increasing productivity. How? When you think of increasing productivity, you usually think of designing new systems to make the machines run faster, or delivery routes more efficient, or the procedures more organized. The Supervisor doesn't generally do any of these things. The Supervisor manages people to take the best advantage of the equipment and systems. And that step is the most critical part of achieving increased productivity.

Sometimes, of course, the Supervisor does have direct impact on improving machines, routes or how to deliver services. Even when that's the case, the Supervisor implements all of this through the management of people.

If you think about it this way, the Supervisor at a fast-food restaurant, the Supervisor on an auto assembly line, the Supervisor of computer technicians, the Supervisor in a pharmaceutical lab, all working in vastly different environments, really have a lot in common: their primary job is to directly manage the people who produce the goods and services that make organizations run.

Leadership and Safe Work Practices

This book is also about leadership, the leadership that must be provided by the Supervisor. Leadership can be demonstrated in many ways. It can come through training, example and motivation. Safe work environments, for example, don't happen by themselves. They require leadership to put them in place and keep them in place. That leadership must start at the very top of a company with a total commitment to safety, to support of the Supervisor in enforcing company rules, policies and procedures, and to recognition of the critical leadership role of the Supervisor. But that commitment is implemented by front-line Supervisors. As in so many other ways, Supervisors directly influence the workplace through their attention to safety.

The Keys To Successful Supervision

Out in Front will also deal with the skills a Supervisor must develop in order to manage people. These include communication, team building, delegating, effective use of discipline, appraising performance, developing employee's skills, motivating employees, problem solving and conflict resolution (Figure 1 – 1). When you look over that list, it should be pretty clear that successful supervision is a learning process, not a skill you're born with.

Dealing Successfully with People

Finally, this book is about successfully dealing with the people you must supervise (with a lot of helpful information on dealing with your managers). Summing it up, you've got to know the job and the situation; your staff—their skills, problems, motivators; and how to effectively handle people situations in the workplace. Let's end with an example of good supervision and a discussion of why it is good.

Figure 1 – 1. The Keys to Successful Supervision

Communication

Team building

Delegating

Effective use of discipline

Appraising performance

Developing employees' skills

Motivating employees

Solving problems/conflict resolution

Integrating safety into the job

CASE STUDY: GOOD SUPERVISION

The Place: A large stamp press and machine shop

The People: Bill, the Supervisor, and Sam, the pressman

The Job: A pressman operates a stamp press machine. He inserts blank sheets of rolled metal (sheets or blanks) into the stamp press machine and then the machine stamps formed products such as trays, washers, coasters, etc.

The Questions: How did Bill handle this situation? Would you have done it differently?

Even before Bill, the Supervisor, was within speaking distance of press operator, Sam Jones, he could see that Sam was really upset about something: he was mumbling to himself and there was a tenseness about him. Sam saw him coming and before Bill could say a word, he barked: *Give this job to someone else, Bill. I'm fed up and want a change! Put me on something else.*

What seems wrong, Sam? said Bill.

Wrong! Maybe something is—maybe it's not. All I know is I've had it! I just don't like this press job. So, move me to something else.

I know how you must feel, Sam. But you must have some idea of what's bugging you. You've done well up to now.

Bill, I told you. I'm fed up. I've been on this thing too long. I can't make time any more and being on piecework, my pay is down. As I see it, I need a change. So, let's just get me a new assignment, OK?

Sam was one of Bill's best pressmen and, at that time, there just were no other jobs available without making a whole series of moves. Anyway, because this wasn't like Sam, Bill wondered what was behind the change in attitude.

Sam, run a half a dozen through for me, will you? Just so I can see if I can find anything, said Bill.

Well, OK, if you say so.

As Bill watched, it was quite evident that Sam was not

up to his usual speed. Then he noticed that Sam fumbled and had to move each piece back and forth a little and try it a couple of times before it would slide into place in the press. When Sam had finished the half dozen pieces, he shut down the press, turned to Bill, and groaned, *See? I'm just stale. Too long on this thing. Did you notice the way I was fumbling?*

I did, said Bill. *It isn't like you either.*

There was a pause as a number of things raced through Bill's mind. Sam seemed excited, probably naturally so, but other than that, he seemed his usual self. Could it be the pieces themselves? That didn't seem likely. They were stamped out on a precision machine and quality control should have caught any errors. After some thought, Bill asked, *Give me a few of those blanks, will you, Sam? I want to check them over. Take a break until I get back.*

Although it seemed of little use, Bill took the blanks to quality control and had them checked. He watched. They checked out a few thousandths off, just enough so that sliding them into place in the press would be difficult. They would have to be lined up exactly, and even a good worker like Sam would have to make several tries. Bill and the quality control person saw the foreman of the group. They checked out the machines and made the necessary adjustment.

When Bill got back, Sam was waiting for him. Sam asked impatiently, *Everything was OK, I guess, Bill, Huh? How about that other job?*

Everything about you is OK, Sam. It's the blanks that were off. They're only a few thousandths off, but it's no wonder you had to horse them around. Send the ones you've got back and the next batch will be OK, that's for sure.

OK, OK, Bill. But I still think I want to be changed from this job.

Bill knew that correcting the errors in the blank sheets had not completely solved the problem. There was a people problem here, too.

Let me finish talking, Sam. I'm going to have all of the time/piecework cards for the past week or so checked closely and starting with the ones where your rate is down, I'm going to adjust all of them back to your previous average and fix it so that you'll get the difference in pay. Can't you see? It wasn't your fault at all. It was the blanks that were out. Anyway, Sam, you really are one of my best men on this job. I'd hate to see you leave it.

Adjusting my pay will help. I'll give it another try if you say so. But if this next batch is bad, believe me, I've had it.

Well, me too, Sam. But believe me, things will shape up.

Now go back to the original questions we posed: How did Bill handle this situation? Would you have done it differently?

Was the problem solved? Partly. Production is back on schedule, but Sam still is thinking about a different job. Bill still has some more work to do if he wants to keep Sam on that job. But that is what supervision is often about: unfinished business. A good Supervisor understands that this is usually the case, finds some immediate solutions to the most pressing problems and then keeps working on that unfinished business.

Could you solve the problem better? You're the only one who can answer that (Figure 1 – 2).

CONCLUSION

In this introductory chapter we have explored the differences between Supervisors and managers and established the fact that the Supervisor is very much focused on the day-in, day-out problems of the workplace. We reviewed the purpose and coverage of this book, and the fact that the focus is on the human resource management job of the Supervisor. We discussed the fact that most of the work of the Supervisor involves people or human resources. This fact follows naturally from the Supervisor's responsibility of directly dealing with all of the organization's production or service resources. The Supervisor is the front-line manager of human resources! That function is a critical element in increasing production and productivity. Finally, we explored the Supervisor's role in providing leadership and safe work practices and the importance of understanding that job.

Figure 1 – 2. How Would You Have Solved This Problem?

Chapter 2

The Front-Line Roles of the Supervisor

Figure 2 – 1. Roles of the Supervisor

Every good Supervisor must be a:

- Manager of the Workplace
- Keeper of Rules and Procedures
- Maintainer of a Safe and Productive Workplace
- Trainer of Employees
- Advocate for the Workforce
- Representative of the Organization
- Leader Out in Front

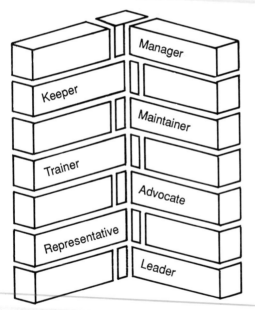

Figure 2 – 2. The Building Blocks of Supervision

OVERVIEW

In this chapter we discuss the fact that the Supervisor is the critical point-of-contact between the organization and individual employees. We also explore the very wide range of activities that form the duties of a Supervisor. These are described as seven key roles that Supervisors carry out in their human resource management function.

The Supervisor is often referred to as being "on the front line," or "on the firing line," or *"Out in Front,"* as we've done in the title of this book. What all of these imply is that the Supervisor is the point for the company. In military terms, the point is the person at the very front of a squadron, probing into new territory. He's the first to see the enemy, the first to fall in a trap, the first to step on a mine. Not a very encouraging concept, is it?

Things aren't quite that bad for the Supervisor in most organizations! However, if there is trouble in the workplace, who will be the first representative of management to deal with it? The Supervisor. If there is a squabble among employees, who usually has to calm things down? The Supervisor. If people are goofing off, who has to get them back to work? The Supervisor, the point. The Supervisor is Out in Front, the person most directly representing the management and the company in its dealings with its employees.

The Supervisor, as the person Out in Front, plays a lot of different roles in carrying out the everyday duties of the job. Let's examine some of these front-line roles of the Supervisor. Each Supervisor, at one time or another, must be (1) manager of the workplace, (2) keeper of rules and procedures, (3) maintainer of a safe and productive workplace, (4) trainer of employees, (5) advocate for the workforce, (6) representative of the organization and (7) a leader, Out in Front (Figure 2 – 1).

Each of these, manager, keeper, maintainer, trainer, advocate, representative, leader, is a key role of a Supervisor (Figure 2 – 2). Like a stack of building blocks, if you pull one out, that is, if you don't carry out the role involved, you can make a mess of the whole job (Figure 2 – 3). Obviously, each role is important. But what is it that the Supervisor is supposed to do?

MANAGER OF THE WORKPLACE

Company presidents think of themselves as impacting on everyone who works for the company. On a day-to-day basis, however, their actual influence is felt by their immediate staff and vice presidents. Vice presidents seem to impact on assistant vice presidents, who in turn reach down to directors or managers, and so on.

Who, then, directly affects employees? The Supervisor, of course. A good Supervisor can inspire, help motivate, encourage safe work habits and procedures and make sure that the job gets done properly and safely. Ultimately, this means that Supervisors affect employees, who ultimately affect productivity and profitability, because it is at the employee level that things are manufactured, services are delivered, products produced. If there is to be profitability in a company, it must be based on the output of the human capital and equipment. It's the Supervisor's job to see that those two elements, human capital and equipment, mesh together (Figure 2 – 4).

Supervisors also have a great deal of impact on the environment in which the work is carried out. Good Supervisors are at the heart of every good company. If the workplace is dirty and disorganized, if things are unsafe or falling apart, you can try to blame it on the lack of money, the failure to buy new equipment or poorly trained workers. Most people agree, however, that the primary blame should be placed on bad supervision. Why? Because the Supervisor is responsible for the direct management of the workplace, getting things done with the people and equipment available. It makes no difference if that takes place in an office, factory, lab or railroad yard, the responsibility of managing is the Supervisor's.

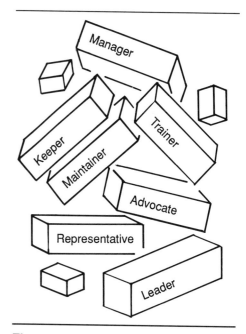

Figure 2 – 3. Making a Mess of It

Figure 2 – 4. Manager of the workplace

If that seems unfair, think for a moment of a small machine shop with fairly old equipment. Some of them make money, some seem to barely exist. Odds are that the successful ones have supervision that keeps tools in good shape, encourages safe practices and keeps production moving in an orderly way.

Ultimately, the Supervisor represents the mission of the organization to the workforce. What is the company all about? What are its purposes, goals and commitments to its customers and employees? All of this is interpreted to the workforce by the Supervisor. The Supervisor is, truly, the Manager of the Workplace.

KEEPER OF RULES AND PROCEDURES

Since the Supervisor is the Manager of the Workplace, it follows that another role must be that of keeper of the rules and procedures. Keeper seems like a funny term, but it's actually right on target. The Supervisor doesn't make the rules, management does that. The Supervisor's job is to make sure everyone keeps to the rules—thus, the Keeper of the Rules (Figure 2 – 5).

Every organization has some rules and procedures that describe how it wants things handled, the objectives of the job, descriptions of duties, acceptable and unacceptable behaviors, the range of behavior between unacceptable and acceptable, the "way things are to be done around here," and so on.

Sometimes, these rules and procedures are set forth in detailed personnel manuals, policy handbooks or labor union agreements. Job sheets, work orders and work assignments also set forth basic rules and procedures. It is important that the Supervisor:

- know what the rules and procedures are
- understand their purpose
- communicate the rules and reasons to employees

As the Keeper of the Rules and Procedures, the Supervisor is both the source of knowledge on the rules and the enforcer of those rules. When there is a labor agreement, the Supervisor should know the contents of that agreement at least as well as the union shop steward. In a sense, both have some of the same job duties.

- Both the Supervisor and the shop steward are obligated to see to it that the union agreement, the mutually accepted terms and conditions of employment, are met.
- Both the Supervisor and the shop steward should know the rules of the labor agreement. If either is ignorant of the rules, trouble will develop.

Perhaps the most frequent cause of formal grievances under a labor agreement is the failure on the part of the steward, employee or Supervisor to understand the rules of the labor agreement. If you, as a Supervisor, take the position that the labor agreement is none of your business, you're in for trouble, and you may find the shop steward running your workplace. On the other hand, if you take the time to learn the agreement and how to apply

Figure 2 – 5. Keeper of Rules and Procedures

it, you will have far less difficulty in dealing with the union and your employees.

The role of the Supervisor is to communicate rules and procedures to staff and make sure they are understood. We will get into the question of communication in more detail in Chapter 4, Communication: A Two-Way Street. The critical thing to remember here is that the obligation to communicate rests first and foremost with the Supervisor. You must start the process, and then you must be sure that communication is always a two-way street (Figure 2 – 6):

- communicating to employees
- listening to what they have to say

The Supervisor must observe workers to see if rules and procedures are being followed. Just because you made the rules clear through effective communication and careful listening doesn't mean that employees will automatically follow those rules. Sometimes they simply forget. Other times they try to find shortcuts around inconvenient rules and procedures—one of the most common causes of safety violations! Therefore, handing out the rules and then sitting back in the office is no way to supervise. You must be out on the floor, in the lab or at the loading dock to observe those whom you supervise. This is a constant process. Never let your paperwork take you away from regular supervision. Direct and frequent observation of procedures and inspection of work is the only way you will ever know if the rules are really being followed. Your employees will respect you for doing your job.

Finally, as the Keeper of the Rules and Procedures, it is the Supervisor's role to reinforce adherence to the rules and to correct employees who are not in compliance. For many Supervisors, both of these things seem difficult to do. Supervisors might tend to avoid confrontation with employees when objectionable behavior is observed. They might prefer to wish away the problem rather than to confront it. That doesn't work. What works is to immediately confront the problem and suggest an immediate remedy. There will be more on this in Chapter 4, Communication: A Two-Way Street.

Praising good or appropriate behavior is more often neglected than carried out. Yet, giving of praise has been repeatedly shown to be the most effective motivator available in the workplace (even better than a wage increase!). This is covered in a how-to fashion in Chapter 4. For now, the two important things to remember are to immediately confront and correct inappropriate behavior or work and to just as immediately praise good or appropriate behavior or work.

MAINTAINER OF A SAFE AND PRODUCTIVE WORKPLACE

Being the Keeper of the Rules and Procedures, you almost automatically become the Maintainer of a Safe Workplace, since safety is usually a matter of following simple but important rules and

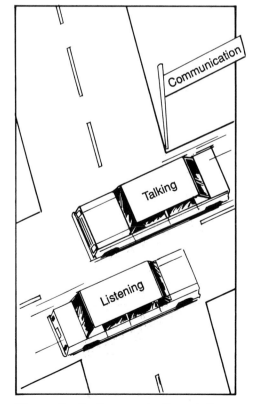

Figure 2 – 6. Communication: A Two-Way Street

procedures and applying a little common sense instead of taking needless risks. Let's look at an example:

CASE STUDY: MAINTAINING SAFETY

A foreman had gone to see how an electrician was getting along with an emergency job. Just as he approached, the electrician, working on a ladder to make a soldered connection, flipped some excess solder from the soldering copper instead of wiping it off. Fortunately, it missed his helper.

The helper was standing looking up, ready to hand him another copper sleeve. Neither man had his safety glasses on. The foreman stopped the work and had the men put on their safety glasses. The foreman stressed the requirement to use safety glasses on the job at all times. He commented, *When you're under pressure in emergency jobs like this, safety glasses can be easy to forget, but they are the best eye insurance you have.* He checked several times during the rest of the day to see if they were wearing their safety glasses.

The direct cause for an accident in this case would have been the failure of the employees to follow the correct and safe working practice. However, an accident may have been prevented by the Supervisor's intervention. Of special interest is the supervisory style of the foreman. He was neutral, nonaccusatory and simply dealt with the facts—an adult working with other adults to improve safety. Worker safety is a prime measure of Supervisor effectiveness.

As this example illustrates, the Supervisor must prevent accidents as well as maintain a safe workplace. The Supervisor must also be aware of observable human factors that can cause accidents, such as fatigue, substance abuse, stress or preoccupation with something other than the task at hand (Figure 2 – 7).

The Supervisor must also know of hazardous conditions in the workplace to correct them, or if that is not possible, to warn employees and establish adequate precautions. Sometimes having hazardous materials around is unavoidable. If that's the case, label them, make sure the employees are aware of the danger and make sure that only those authorized to work with or move the materials do so.

As the Maintainer of a Safe and Productive Workplace, you must also be familiar with special requirements of hazardous materials, storage regulations, the Right-to-Know Act, the general provisions of the Occupational Safety and Health Act, and first aid procedures. This book does not attempt to cover all of these details, although Chapter 7, Supervisors and the Law: Employment and Workplace Law, will deal with many of the legal aspects that must be understood by Supervisors.

Figure 2 – 7. "Human Factor" Warning Signs

TRAINER OF EMPLOYEES

Supervisors have to do a lot of training. Some of it is direct, such as sitting next to a person at a terminal and making sure he/she knows the basics of the computer software. Another kind of training can be what is called modeling. Yes, people do see how you act, how you move, how you handle your job. Good or bad, you're serving as a model for them and training them by your actions. So, you must set a good example.

Some Supervisors think that training employees is not a part of their job. This is the sink or swim school of supervision. If you follow it, most employees sink. The Supervisor must provide proper training so employees can succeed.

The most common ways Supervisors provide training in the workplace are:

- Orienting and instructing new employees
- Setting a good example to follow
- Allowing employees to practice new skills
- Observing employee skills and providing positive feedback or corrections

Supervisor-provided training also includes orienting an employee to the goals and standards of the organization, the purpose of the employee's job and the performance expectations in that job. Supervisors should emphasize to new employees that safety is a part of the job and will be a factor in performance evaluation.

ADVOCATE FOR THE WORKFORCE

Because they know the needs of their employees better than anyone else in management, Supervisors often find themselves serving as an Advocate for the Workforce. Don't be alarmed at how often you seem to be the one complaining about the employee cleanup room, the lunchroom offerings, or the fact that the workforce is unhappy with their wages. Because you are trying your best to get production out of people, you are well aware of the workplace factors that pull that production down, and you naturally want to change things.

The Supervisor actually acts as a liaison or communication link between upper management and the workforce. This is a critical link, and without it, the company wouldn't function. The Supervisor is the representative to management of the workforce and their needs. Supervisors must insure that employees have the proper resources (i.e., tools, equipment, safety and personal protective equipment) to perform productively and safely.

Being an advocate means being attuned to employee needs. This does not mean you're a mind reader. It is better not to assume you know what your employees want, but to listen, ask and gather information from employees.

Obviously, you must selectively express employee concerns with management. (You're not their union representative!) You will often find that there is no immediate solution to some problems. You might also find that management is tired of hearing the same story. That doesn't mean the problem has disappeared, or that you should give up if you believe the issue is very important.

The best approach is to pull back a bit and do a little research on the issue involved. If it's a change of job duties, for example, then check into other similar jobs and make some comparisons. Present a tighter, better researched argument next time there is a good opportunity.

Sometimes, of course, you do have to back off. Management might not always be ready to make certain changes, even if you believe the changes to be good ones. Knowing your organization and when the time is right is also part of being a good Supervisor.

REPRESENTATIVE OF THE ORGANIZATION

Although the Supervisor is clearly the Advocate for the Workforce, Supervisors are also expected to carry out the role of Representative of the Organization (company, management). They must stand for and be able to communicate the company's objectives, priorities, expectations, concerns and even appreciation. Employees look to their Supervisor for guidance on what the company really is doing, what it wants its employees to do and what it hopes to achieve through all of this.

Company policies must be the Supervisor's policies, or ultimately employees will lose respect for the Supervisor's authority. Taking a position such as *Well, that's what they told us to do, but I think we can do it my way instead,* means you, the Supervisor, do not respect company policies. You will ultimately be at odds with your manager, and you will lose. When you lose like that, you also lose the respect of those you supervise.

As Representative of the Organization, it is your responsibility to make the company resources known and available to employees. If there is an eye clinic service provided, let your people know about it. Be familiar with the medical and health benefits. Know what kind of counseling is available for various personal problems such as substance abuse, marital problems, child care and so on. To your employees, you are the Representative of the Organization and their personal connection with that organization.

Of course, you must also keep in mind that the Supervisor is management, the company, in the eyes of many employees. It is always important to set an example that expresses your relationship.

LEADER OUT IN FRONT

Finally, you are the Leader of the workforce you supervise. You are truly a Leader Out in Front because you must deal directly with the part of the workforce that actually makes the things the company produces, or provides the services the organization delivers. Those above you are managers. They are not out in front in terms of the relationships with the majority of the workers they manage. You are.

You become the focal point of the workplace. You direct, guide, encourage, reward, punish, counsel and advise those who work for you. It is a very personal relationship, one that is not likely to be held by your manager or the company president. You must provide a vision to those who work for you in order to focus their talents and energies on carrying out safe and productive work.

It's a tough job. Who ever said it would be easy? On the other hand, being a Supervisor is also a very rewarding job!

CONCLUSION

In Chapter 1, we discussed what a Supervisor is and how that is different from the job of a manager. Building on that information, in this chapter we have discussed the fact that the Supervisor is the company's primary point of contact with the workforce, the point person of the company.

We have reviewed the many roles a Supervisor must carry out as a routine part of his/her job and have found that it is quite a task. It seems that you're required to move back and forth, from one role to another, almost like changing uniforms and hats. Yes, there are many hats that a Supervisor must wear, many roles to play in the workplace. These seven mentioned here are the critical ones, and you should always be conscious of them. Look them over once again (Figure 2 – 8).

Figure 2 – 8. Roles of the Supervisor

Every good Supervisor must be a:

- Manager of the Workplace
- Keeper of Rules and Procedures
- Maintainer of a Safe and Productive Workplace
- Trainer of Employees
- Advocate for the Workforce
- Representative of the Organization
- Leader Out in Front

Chapter 3

Leadership: The Evolving Job of the Supervisor

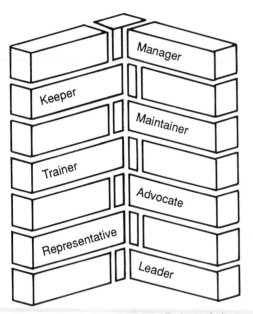

Figure 3 – 1. Interlocking Roles of the Supervisor

Figure 3 – 2. Whose Team?

OVERVIEW

This chapter deals with how you exercise supervisory leadership in the workplace. It further explores your role in management, and then moves on to the type of supervisory leadership styles available to you. It explores the relationship between supervisory leadership and motivation of those who work for you, and ultimately suggests a flexible supervisory leadership style that will aid you in dealing with the differing abilities of those who work for you.

PART OF THE TEAM—BUT WHOSE?

When you think back to Chapter 2 and the various functions carried out by the Supervisor, as illustrated in Figure 3 – 1, it's pretty clear that you've got a lot of roles wrapped into a single job. It's also clear that there are some major contradictions or tensions between those roles. Most of these relate to where you fit: with the workers or with management. In fact, it often seems that the Supervisor is mainly doing a balancing act between representing the organization and being an advocate for employees (Figure 3 – 2).

Since most Supervisors are promoted from the ranks, they probably have many friends in the ranks and very few friends in management. Can you keep those friendships? Probably, but sometimes it will get difficult. The fact is, you can't be a Supervisor and be just one of the gals and guys at the same time. Management won't let you do that, nor will your former fellow workers!

You have to do a balancing act. A Supervisor is management. But a Supervisor also happens to be standing on the first rung of the management ladder, closest to the workers. Putting it simply, you must keep both facts in mind to be a successful Supervisor. At the same time, you must lead. That is the heart of your responsibilities.

While the employees you supervise may resent your new position, or tease you about it, they do understand that a Supervisor is there to lead, and that makes you a leader and a member of management. Supervisors who try to run away from that idea and pretend that nothing changed when they were promoted usually do not succeed. On the other hand, Supervisors who let it all go to their head and mistake the position of Supervisor for president have a really terrible time of it. Again, balance is required. You will have to earn the respect of your employees, and this balancing act helps you to achieve this respect. Title alone does not make the leader.

What kind of leadership does a Supervisor really provide? In Chapter 2 we discussed the roles you play, most of which are related to leadership, as shown in Figure 3 – 3. Now, we will look at the functional areas of your job that really define your leadership in the workplace. The functions we're talking about have to do with the very nature of the Supervisor's job: what it is Supervisors do that is inherently different from what workers do and thereby expresses the leadership authority of the Supervisor.

Components of a Supervisor's Job

The five main components that define supervisory responsibility and authority are as follows:

First, Supervisors define, or help define, the work to be performed and the goals to be achieved by the employees they supervise. This function of planning and setting objectives is critical to success. If you do not do this, or you do not have the authority to do this, you will not be accepted as a leader.

Second, Supervisors have responsibility and authority over the use of at least some of the resources needed to do the work. Supervisors, of course, never feel they have quite as many resources as they would like. Nonetheless, they must have some resources, and the fact that they exercise such control establishes them in a leadership position. Control of resources, even of a limited nature, has always demonstrated power, and power is an element of leadership. If you doubt this, think about the supply sergeant in the army, or the clerk who controls the office supply room. You may be annoyed with their power, but they do have it.

Third, Supervisors represent the interests of their employees to others in management. Employees look up to whoever looks out for them. If you're an effective Supervisor, that means YOU. If you do not do this, employees will not see you as their supporter, which will diminish your leadership potential.

Fourth, Supervisors resolve conflicts in the workplace among their employees. If you walk away from or ignore conflict, you're letting everyone down and they'll let you know it. If you use this authority, some employees may not like your decisions, but they will respect your exercise of leadership.

Fifth, Supervisors set an example for their work group in how they handle themselves in carrying out these activities. If you are arrogant, arbitrary or capricious in your exercise of authority, you may still be accepted, grudgingly, as a leader. If you are wise, consultative, helpful and supportive in your exercise of authority, you will be respected, admired and followed.

All of these functional areas of supervisory authority represent things you can and should do. They also show the difference between a Supervisor and an employee.

Management's Team

At the beginning of this chapter we stated that the Supervisor is part of the team—but whose? If there is still any question in your

Figure 3 – 3. What Areas Involve Supervisory Leadership?

1. Supervisors define, or help define, the work to be performed and the goals to be achieved by the employees being supervised.

2. Supervisors have responsibility and authority over the use of at least some of the resources needed to do the work.

3. Supervisors represent the interests of their employees to others in management.

4. Supervisors resolve conflicts in the workplace among their employees.

5. Supervisor set an example for their work group in how they handle themselves in carrying out these activities.

mind, the answer is management's team. You're on that team because you have responsibility and authority that have far more to do with management than with your former fellow employees.

We are now going to turn to four major components that you need to examine in order to determine an appropriate supervisory leadership style that will fit your needs and assist you in becoming a more effective leader:

- Leadership styles
- Who you are leading and what motivates them
- Adjusting your style to the need
- Finding the right ways

SUPERVISORY LEADERSHIP STYLES

There are many ways in which Supervisors can express their leadership. Some styles don't work very well, but others do. Generally, they range from an authoritarian approach to a very democratic approach. We will discuss many aspects of leadership style later in this chapter and define the various styles. First, let's look at three examples of different leadership styles used by three different Supervisors dealing with the same problem. Our purpose here is to see that, in fact, there is a difference in leadership style (whether you give it a label or not), and it does produce different results. We will then look at some of the finer distinctions.

CASE STUDY: LEADERSHIP STYLES

Three Supervisors all have the same problem. The following cases show how these very different Supervisors handled the problem. Each Supervisor has a production bottleneck. Each has found that it is due to the disorganized way in which the employees are arranging and storing their completed work pending its periodic pickup.

CASE 1: SUPERVISOR SUE PERKINS

Supervisor Sue Perkins gets the four men concerned together near their machines and says, *My checkup shows we have a production slowdown on our hands. The way you're handling and storing boxes of completed parts is actually cluttering things up. The aisles and passageways around your machines are a mess. Also, there is a lot of odd scrap lying around. By tomorrow noon, I want the machine benches cleared of everything but tools and work in progress. I want completed parts stored free of the walkway and all of the scrap sorted and disposed of. I'll make a check on it first thing after lunch tomorrow.*

OK, Ms. Perkins, say all four men.

Sue Perkins' orders are carried out without delay. It results in an easier flow of material, a noticeable improvement in the department's appearance and, at the end of a week, a rise in the output.

CASE 2: SUPERVISOR HUGO SCHMIDT

Hugo Schmidt has a different way of approaching his production problem. *We've got to fix this production falloff that we're having. Let's give it a little thought right now. I'm open to suggestions. What do you think?*

We've got to have things within reaching distance if we're going to work at top speed, said Hal.

Why do you say that Hal?, Hugo asked.

Well, I thought maybe you felt things were sort of messed up and disorderly.

What do you think about it, Bob?, asked Mr. Schmidt.

Looking around, I guess we have gotten a little careless and scrap hasn't been taken away and we've just been stacking the boxes of finished parts any old way until they're picked up later in the day.

Looks to me as though you've put your finger on something, Bob. How does it seem to you, Tom?

Well, Hugo, as I look around, I've got to agree with Bob.

It seems to me you fellows have found at least one of the sore spots. It shouldn't take long to clear it up, though, said Hugo.

Hugo, I think we could clear this up by tomorrow night by doing a little at a time between runs, said Bob. *How about it?*

We can do it, the men agreed.

Sounds good to me. I'll check with you late tomorrow afternoon and see how it's going. If you get stuck on anything and need any help from me, let me know.

By the next afternoon, things were in good order and when Hugo Schmidt came around, he spoke to each of the men about his own machine and work area, complimenting each on a good job. Things began to run more smoothly right away, and in ten days' time, production was on the rise.

CASE 3: SUPERVISOR MICHELLE POWERS

Michelle Powers handles it quite differently from the others.

We've got a problem about production. It's fallen off a lot. I wondered if you don't agree that things are too cluttered, and we need to straighten up and get a better way to store the completed work?

Maybe you're right, boss, said Tom.

Could be, said Jim.

Well, take a look at it, and see if there isn't something you can do, said Michelle.

From time to time, the group made a feeble effort to clear out around the machines and get rid of the scrap, but most of their effort consisted of moving things from one spot to another. The production falloff lingers on and on with no improvement.

Each of the three Supervisors have leadership patterns that are characteristic of a different leadership style. In the first case, Sue Perkins used a dictatorial or autocratic approach characteristic of leaders who determine all by themselves what is to be done, how it is to be done and when it is to be done, and then assign tasks on an individual basis. They set deadline times, too, and then check on schedules.

Hugo Schmidt got the men in on the problem and its solution. He encouraged the men to participate in the decision, was direct and objective in his comments and checked up on the job afterwards, giving credit where credit was due.

The third Supervisor, Michelle Powers, did not exert control. She pointed out the problem, left it up to the men to do something about it and then just let things ride along.

Two of the leadership styles, although very different, resulted in the needed improvement of production. One did not. Clearly, we can see that Michelle Powers' approach lacked leadership. Hugo Schmidt and Sue Perkins, on the other hand, were clearly leading. What was the difference between them, and which approach was best in the circumstances involved?

FACTORS IN LEADERSHIP

Leadership styles really run across a wide range of supervisory behavior. Generally, they show a difference between leading and managing. At their extremes, leading often means being Out in Front, pointing the way, and setting a good example for others to follow. Managing, on the other hand, means delegating responsibility, assigning tasks and waiting to see what happens. Which style works? Both approaches can be effective, and both have their limits. The good Supervisor learns to mix the functions of leading and managing. Someone who does only one and not the other will not be an effective Supervisor.

Other factors of leadership include motivating people as opposed to controlling them and inspiring employees versus guiding them. A Supervisor, once again, needs to mix all of these factors in his/her leadership style (Figure 3 – 4).

You must learn to motivate those who work for you. Motivation techniques should be different for different people and situations. Sometimes they are inspirational: talking about the team, the need to get the job done, the organization's mission or the importance of a safe work environment. Other times they must be controlling: You simply must do things in a certain way. The motivation to keep a safety shield in place is to prevent injury to people. However, if an employee has not seen anyone get hurt when the shield was off, that motivation may not work. The Supervisor needs to point out near misses and use a straightforward controlling policy. The employee must be told, *You will be penalized if you do not follow the rules and keep the shield on.*

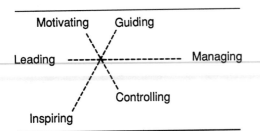

Figure 3 – 4. Factors of Leadership

There are many times when a Supervisor must get things done quickly. In those cases, there is often no time for motivation and explanation, or delegation of responsibility. It is, instead, a time for controlling. The company has, after all, placed you in control. You must learn to make the call as to how much control has to be exercised at any given time and under differing circumstances.

Inspiring people versus simply guiding your work team is yet another set of leadership approaches. Inspiration through example, deeds and words is important. You spur people forward. However, too much inspiration, like all good things, can wear very thin. In fact, people seem to respond more effectively to guidance than to inspiration. Guiding is a lower-keyed, lower-pressured form of leadership.

For example, if you're having difficulty meeting a schedule, and you go to your manager to discuss it, which of the following responses by your manager would work best with you?

1. *Well, now I know that schedule's a bit tight, but you're the best Supervisor and crew we've got, and I'm sure that when you put your heads together you'll figure it out and bring it all in on time!*

OR

2. *If you think it's too tight a schedule, maybe we need to look at delaying delivery so you're sure we can do it right. Before that, though, why don't you and your crew take a look at the project again and see if there aren't some ways we can improve production enough to meet or come close to the schedule. Take a little time at lunch today and talk it over, and then let's have another look at it. I'm really interested in seeing what you can come up with, but understand that if things have to be delayed, they will be delayed.*

If you're like us, you prefer the second approach. Yet, both approaches are really saying something very similar: Go back to your crew and work things out. Approach 2 makes us feel a lot better about it, and also lets us know it's not sink or swim.

All of these leadership factors are important, and as usual, maintaining a balance between them is ideal. You will develop your own personal mixture.

Four Leadership Styles

Now that we have looked at some of the variables involved in leadership—leading versus managing, motivating versus controlling, inspiring versus guiding—let's put some labels on the various styles that fall between the extremes. There are, probably, an innumerable combination of supervisory leadership styles, but the main ones most people in human resource management deal with are:

- Authoritarian Supervisory Leadership
- Democratic Supervisory Leadership
- Hands-Off Supervisory Leadership
- Flexible Supervisory Leadership

Authoritarian Supervisory Leadership

Authoritarian or directive supervisory leadership is often characterized as old-fashioned. In many circumstances, it is not only out-of-date but ineffective. Authoritarian leadership involves giving orders, and leaving little or no room for input or questioning by employees. The authoritarian Supervisor usually:

- sets specific time-frames by which things are to be completed,
- lays out all work expectations,
- goes over all the how-to's and makes sure they are understood,
- retains control by checking on the employee frequently to assure the expected outcome.

Authoritarian leadership may be appropriate with some employees, such as newly hired employees until they have learned their job responsibilities and company procedures, employees placed in positions they are not really trained or prepared to handle, employees handling new and unfamiliar equipment, employees who don't want to have to think but only follow orders, the unskilled or untrained employees hired on a temporary basis or temporary employees in general. (They may not need it, but you won't have time to learn that.) While it may be old-fashioned in many parts of the workplace, authoritarian supervisory leadership still fits in these circumstances.

For example, if you are short-handed and are sent a new employee who is supposed to handle an expensive and dangerous piece of equipment, you had better use an authoritarian approach: set specific time frames for each task, lay out all work expectations, go over all the how-to's regarding the equipment, safety rules and the job, make sure they are understood and check frequently to assure the expected outcome.

Authoritarian leadership is usually not appropriate with employees who are highly skilled or educated, long-term employees, creative or innovative employees. It tends to set them on edge, disturb them and destroy their creativity and productivity. Even with these people, however, it may have to be used in emergency situations if time doesn't allow for input from others. If a fire starts, you don't debate it, you don't form a committee to discuss fire control. You give clear, precise and authoritarian orders to put out the fire.

Having established that authoritarian behavior is sometimes necessary, it is important to stress that it is not necessary most of the time and is usually counterproductive for the reasons already stated. Unfortunately, authoritarian behavior is often what Supervisors fall back on when they feel insecure or threatened. It's an old role model—the stern grade school teacher, the angry parent—and even though it usually didn't work, it's somehow still there in people's memory banks. The fact is, if you're feeling threatened or insecure and use authoritarian behavior to cover up, it usually won't work. Why? Because you're feeling threatened or insecure, it will show.

Democratic Supervisory Leadership

Democratic or nondirective, participative supervisory leadership in all its various forms from Japanese Style to Modern American Management Style is usually seen by human resource managers as the appropriate way to supervise. Under the democratic approach, employees provide input into how the task or project is to be completed. That is, they participate in the decision-making process.

This is usually called "delegating." When a Supervisor delegates, that means a portion of the Supervisor's job is actually being assigned to an employee. Thus, the Supervisor delegates some of her/his authority to the employee to do the job, assigns responsibility to the employee to make the decisions involved, while always retaining overall responsibility. Responsibility, in this way, is actually "shared."

The level of control by the Supervisor in this process can vary. Some Supervisors can allow the employee or employee group to make decisions:

- Totally on their own,
- With some input by the Supervisor, or
- With input from the employees but the final decision retained by the Supervisor.

In any event, while Supervisors delegate responsibility to employees in varying degrees, they still retain full responsibility for the results of the employees' efforts.

Characteristics of the democratic Supervisor include: security, confidence in the staff's ability and competence, recognition of the need for staff development and willingness to share the control. Characteristics of employees who respond well to this style are familiarity with the job; required skills and organizational rules and procedures; a willingness to participate in the decision-making; and a higher level of experience, skills or seniority.

Just as there are problems with an authoritarian style of supervision, there are also problems with the democratic style. Some possible pitfalls include:

- Many employees don't want to participate in the decision-making process. (You don't pay me to do that. It's not in my job description.)
- Many Supervisors really only give the process lip service, letting employees give input but ultimately implementing the Supervisor's original plans anyway.
- Many problems do not need or lend themselves to group decision-making.

Hands-off Supervisory Leadership

The hands-off Supervisor assumes employees have the skills and resources to perform their job and provides no direction or feedback in the process. A laid-back type, the hands-off leader is often seen as the answer to a workforce focused on finding job satisfaction.

The hands-off leader pretends to be a model of delegation. However, the reality is that both authority and responsibility for tasks are being delegated, and the hands-off Supervisor is trying to walk away from responsibility. That simply doesn't work. The Supervisor cannot walk away from ultimate responsibility. Any delegation of responsibility or authorization to an employee to carry out a specific job remains a shared responsibility with the Supervisor, and the Supervisor is ultimately responsible to the organization.

The Supervisor can delegate to an employee by giving him/her specific responsibility for doing part of the Supervisor's job. This delegation conveys the "authority" to carry out specific tasks or functions, and the "responsibility" to make the decisions involved as agreed to. In this way, new responsibility is created for the employee involved, but the Supervisor retains ultimate responsibility for what is or is not accomplished. It is a shared responsibility situation. Both the supervisor and the employee share the risk (and responsibility), although the Supervisor holds ultimate responsibility and accountability in the eyes of the organization. Nonetheless, if you expect the employee to do a job, you must be prepared for such risks.

Another problem is that this style doesn't work in most cases and can as easily be described as the lazy leadership style. There are, of course, some employees who are motivated by feeling they are trusted and respected to perform without much, if any, intervention. Presumably, they would be responsive to this style. We suspect they are few, since most would still feel some frustration about a Supervisor who did little, if any, work.

The primary drawbacks of this style are pretty apparent. The hands-off Supervisor never gives employees any positive reinforcement (Everyone needs a pat on the back from time to time!), and mistakes are usually not corrected until there is a crisis. Many Supervisors of this type see themselves as a resource person whom employees will come to when there is a need. That, of course, assumes the employee will realize that there is a need.

Now go back and look at the case study showing very different styles used by the three Supervisors. In Figure 3 – 5 write the supervisory style used by each.

Figure 3 – 5 . Supervisory Leadership Styles

Supervisor	Supervisory Leadership Style
Sue Perkins	_____
Hugo Smith	_____
Michelle Powers	_____

Flexible Supervisory Leadership

The fourth style, flexible supervisory leadership, is a combination approach. It has been described by many writers, most thoroughly by Kenneth Blanchard and Paul Hershey in several books. Blanchard and Hershey refer to it as "situational leadership" and use the concept in a management as opposed to a supervisory model. That is, they focus on things managers do rather than things Supervisors do.

Flexible supervision means choosing the style that fits both the particular situation and the employee(s) you are working with in that situation. This means the Supervisor must be flexible in the style to be used at a particular time and place and with particular employees.

Since most Supervisors have a style they are most comfortable with, they must first become aware of what their own usual style is before they can become aware of how to use other styles. Look back over the characteristics of the styles we have described. You will probably find yourself somewhere between two of these.

Now comes the hard part: you can bet that one style is not appropriate for all situations. There are work situations where you simply must be authoritarian: enforcement of safety procedures, maintaining firm rules on substance abuse, and so on. Blanchard and Hershey call this style directing. They see a variety of styles ranging from directive to delegating (what we have called authoritarian to democratic). Figure 3 – 6 shows the wide range of styles within the authoritarian to democratic field. The flexible supervisory leadership style can utilize all of these styles, depending on the particular situation and the employee(s) involved!

You must face up to the fact that while you try to keep teamwork going, a team is always made up of different individuals with many different abilities, motivations and needs. Getting to know your employees' individual needs and skill levels will help you determine which employees respond best to which

Figure 3 – 6. The Range of Leadership Styles

```
A                        F L E X I B L E                        D
U    <  - - - - - - - - - - - - - - - - - - - - - - - - - >      E
T                                                               M
H                                                               O
O                                                               C
R  ═══════════════════════════════════════════════════════     R
I     D    H    A    C    S    S    D    H                  A
T     I    E    S    O    U    U    E    A                  T
A     R    L    S    A    G    P    L    N                  I
R     E    P    I    C    G    P    E    D                  C
I     C    I    S    H    E    O    G    S
A     T    N    T    I    S    R    A
N     I    G    I    N    T    T    T
      V         N    G    I    I    I
      E         G         N    N    N
                              G    G    G
```

leadership style. Being flexible enough to recognize that Jim is a fast learner, has the basic skills and is able to handle responsibility very well means that you can delegate a great deal to him—almost becoming a hands-off Supervisor, but certainly a delegating type (and without the laziness!). Carol, however, cannot learn so quickly, is a bit afraid of moving on to new projects, but does have some skills. You're going to have to coach her a lot, in a very supportive but democratic manner. Reggie, on the other hand, waits for orders. *Tell me what to do, and I'll do it*. He may know how to do it, may possess the necessary skills, but he's one of those *You don't pay me to think* types, and the only way to handle him is with very authoritarian instructions.

There's no way to learn this, except practice. Practicing different styles with your employees in different situations will help you as Supervisor reach a balance. That's flexible leadership.

But before you can do that, you must know something about those you are leading, the employees you supervise. You need to learn methods to analyze employee needs before you flexibly respond to their needs.

EMPLOYEE MOTIVATION

It's clear that learning about your employees covers a lot more than what is in their personnel files. The files are records. People are made up of a lot more than records, and when they come to work, they're bringing a lot of past history and personality with them. The first thing you need to look at is what motivates them at work.

What motivates you will probably not be the same as what motivates your employees. Don't assume that what satisfies you will satisfy them. In Japan, many Supervisors and their employees go to lunch together every day and have drinks together every work night. That's a little much for most Americans, but you do have to get to know your employees so you can learn what motivates them. That will take time and attention paid to each individual in your group. What motivates one staff member may not motivate another.

Factors of Motivation

There are many factors that motivate us to work (or not to work). If people aren't making a decent wage, don't expect a lot of work out of them. In the Soviet Union everyone is guaranteed a job. To achieve this, they overstaff most jobs (three or four people assigned to do what one person could do). Most of these people receive very little money working at their guaranteed job, and productivity is very low. Not much to do. Not earning much. Little motivation.

Something a little closer to home and harder to spot is a lack of motivation because of fear for one's safety. For example, if you're working in a place with hazardous waste products, you see them regularly mishandled and you know the dangers, you're not likely to be thinking about your work. You're thinking about protecting your life! No motivation for work, but plenty of anxiety.

Figure 3 – 7 shows the motivating factors regularly mentioned in employee motivation studies. Look them over. If some of these are a problem with your team, try to find a way to address them.

Respect, praise, being part of the team, personal growth, money, recognition, job security, a safe workplace, responsibility, working environment, participation, challenge and autonomy are all seen as factors that help motivate employees. These things are not all the same or equal. And different factors affect people differently at different times.

In fact, some of these items are not quite motivators. Let's look at a different grouping of these motivators.

Demotivators

Money
Job security
Workplace safety
Environment

Motivators

Respect
Praise
Responsibility
Participation
Challenge
Autonomy
Part of the team
Growth

You'll notice we've called the group on the left demotivators, which seems strange since we all tend to value these items very highly. A demotivator is something that will not necessarily make someone work harder; but, if it is absent or lacking, it can cause someone to perform less or at a lower quality level. Ask yourself this: if the company gives all employees regular large wage increases six times per year, will that motivate the employees to work harder? You'll probably answer, *Maybe the first few times, but after that, well....* And you're right. The money isn't a motivator.

On the other hand, if the company regularly pays employees way below local wage standards, can you still motivate employees using the items from the list of motivators? Probably not.

What do we conclude? Money isn't a motivator, but the absence of an appropriate amount of money is certainly a demotivator. It's also easy to see that both motivators and demotivators affect employee behavior, thus motivation. Look at each of the other items in the two lists and think them through: why are some motivators and others demotivators?

We have to remember, of course, that motivating factors, like people, are not static. They change with the time and the situation. What motivates an employee one day may not motivate him or her the next. It's the old *What have you done for me lately?* routine. Not only do you have to get to know your employees and what motivates them, you also have to be aware of changes.

Problem Solving, Motivation and Building on Success

To motivate others, it is often useful to give them a work-related problem to solve. Many people respond to that type of challenge. It seems like a simple matter:

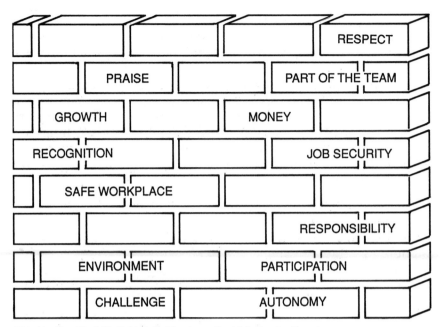

Figure 3 – 7. A Solid Wall: Factors that Motivate Employees

Joe, I'd like you to look at that spool holder and figure out how we can make it operate more smoothly.

But is Joe likely to be able to solve the problem? Does he have the knowledge, skills and background? Does he have the tools? If it's all go, then try it out. This approach is called training people by building on success. If he doesn't have the skills and tools, don't do it. He's doomed to failure.

Building on success is a technique that allows the Supervisor to gradually increase the difficulty of problems he or she can delegate to subordinates. Since delegating work is one of the major ways a Supervisor can work smarter rather than harder (as we will discuss in more detail in Chapters 5, 10 and 13), it makes sense to look closely at these ideas.

There are four key principles involved in developing people through the technique of building on success (Figure 3 – 8).

First, give people problems they can solve. If they don't have the knowledge necessary to figure out the problem, they will fail. Failure, while faced by everyone, is not the most effective way of learning. Success is.

Second, give people the tools to solve the problem. If you have the answers but don't have the tools to implement them, then where are you? More frustration and failure.

Third, give people some praise and acknowledgment for having solved the problem. (If they do not solve it, give them guidance that will encourage and enable them to solve it.) The greatest tool you have available to motivate your employees is praise. Use it generously, but don't use it falsely. At this point, having given them a problem they can solve, given them the tools to implement the solution, and praised them for the results, you have completed a step in the employee development process. Now we move to the fourth principle.

Fourth, continually increase the difficulty of the problems you assign. This means you will be increasing the difficulty of achieving success but always keeping it achievable. This way, you constantly provide a challenging opportunity to experience success and satisfaction, and you share some of your work with those you supervise. All of this will provide job satisfaction and motivation to do better.

Types of Employee Behavior

Knowing your employees also means having some understanding of the basis for their behavior. Employee behavior (like all behavior) is a function of what motivates a particular individual. Employee behavior is also a function of the situation in which the individual is working, the effects of what happened outside the workplace and past beliefs and attitudes.

Many times, in looking at a problem employee, a Supervisor may simply ask himself or herself: *Is Bernie having a "good" day or a "bad" day?* Often, that's enough to give you some immediate insight and understanding. We all have good days and bad days, and we learn to respect this in others. At the same time, you must avoid simply dismissing real problems as just a bad day.

More difficult problems can arise when negative behavior confronts the facts of working life.

- An employee can act resentful or difficult if the Supervisor is a woman, younger than the employee or a member of a minority group.
- Jealousy can be present if the Supervisor has been recently promoted from the ranks.

In cases like this, the employee must be taught that the Supervisor's position is a fact of working life that cannot be adjusted, so the employee must learn to accept reality. It's as if the employee has a filter lens on himself or herself, and sees the work world very differently than you do. When you are aware of the problem (the filter through which the employee is viewing the workplace), you, the Supervisor, can help the employee to accept reality. You do this by expressing clear expectations, demonstrating your own basic work motivation and, as necessary, directly confronting unacceptable behavior.

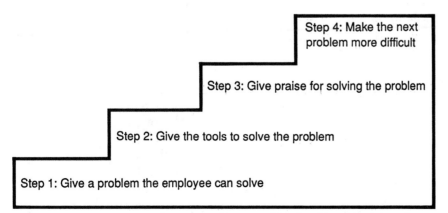

Figure 3 – 8. Building on Success—Steps to Sharing Your Work Load

If the behavioral problems persist, understand that you cannot control the employee's biases or jealousies. You can and should confront and control unacceptable behavior. Focus on the behavior, and make sure the employee knows exactly what is expected. If that expectation is not met, you may have to take disciplinary action.

As we all know, outside forces also create behavior problems in the workforce. An employee can appear fatigued if there are family problems, medical problems, substance abuse problems or even if there was a good late movie on last night. Sometimes a little personal inquiry can identify a problem, bring it out in the open and allow you and your employee to briefly deal with it. This interest on your part can be sufficient to cause a positive change in behavior.

Remember, however, that being a trained counselor, doctor or nurse is not one of your many roles (unless you happen to be a Supervisor in those professions!), so don't push too hard on the questioning. It is always best to stick to the facts of the performance problem rather than diagnosing or recommending solutions to personal problems.

If behavior motivation problems persist, it may be time to suggest the Employee Assistance Program (EAP), if your organization has one, or to talk to your manager about the employee's problem as you see it.

Ultimately, employee behavior is a result of the individual's personality. Some employees will be outgoing and others shy. Some employees will be go-getters and want to move ahead while others will be content to stay in the same position. Some employees will want an organized workstation while others can work effectively in clutter. It is the Supervisor's role to observe employee behavior, reinforce positive behavior and correct inappropriate and unsafe behavior. It is the Supervisor's role to be aware of what motivates individual employees. When Supervisors learn to look at these issues, they can move on to be effective flexible leaders.

ADJUSTING YOUR SUPERVISORY STYLE TO THE NEED

It's clear that the ability to adjust your supervisory leadership style is important to successful leadership. Yet, some difficulties can be created when you try to adjust your supervisory style to the needs, skills and situation of your employees and the work to be performed.

- First, it is difficult because the burden to adjust is placed on you. So what else is new? Leadership has its price, of which you are already aware. The burden is being placed on you, since it is part of your job. The best approach is to go at it slowly but steadily. Developing a flexible supervisory leadership style is like developing any skill: it takes time, practice and a lot of effort.

- Second, it is not always easy to figure out the needs and expectations of the organization in which you function and how far they want you to go with participative management.

Does the organization cultivate a particular management style? If so, they may well expect all Supervisors to follow this particular style. Don't rush to judgment on this score as a way of avoiding the work and difficulty of being a flexible leader. Most organizations, in fact, contain a very wide variety of styles of leadership. Most companies want flexible leadership in their Supervisors and managers, even if they have only a few who practice it effectively.

With a flexible leadership approach you simply have to keep posing questions based on the situation and the members of the team you supervise. Figure 3 – 9 gives you a flexible leadership checklist.

FINDING THE RIGHT WAYS

A flexible supervisory leadership style makes the most sense. It allows you to meet company and individual needs and assures greater opportunity for getting the work done in a timely manner. However, it isn't going to be easy.

Awareness of different styles gives you the first set of tools. It causes you to think about how you have been supervising, and to consider some changes in order to be more effective.

Determining what you are comfortable with is critical. Some people simply can't tolerate a hands-off approach to supervision even if their employees are perfectly capable of carrying it out. If that's the case with you, scale back a bit from hands-off to participatory. Seek the employees' advice and suggestions, but retain the Go-Ahead-With-It decision.

Observing and getting to know your employees is clearly a very important component. No matter what leadership style you are most comfortable with, or what mixture of styles you can use, you won't get very far unless you really know your people.

Finally, practice really does make perfect—or at least close to it! Try out different styles with different employees based on

Figure 3 – 9. Flexible Leadership Checklist

_____ What do your employees need?

_____ What do you as a Supervisor need, and can you effectively and comfortably use different styles?

_____ What are the needs of the situation?

_____ Is it an emergency situation? A directive approach may be necessary to avoid an accident.

_____ Is there time to get input from employees? If so, you may be able to use a participative approach.

_____ Do employees have to feel ownership in the project? If so, a participative approach is probably appropriate.

_____ Is the employee a long-time, highly skilled and respected employee as well as a self-starter? If so, a hands-off approach may be effective.

Table 3 – 1. Comparison of Leadership Styles					
Leadership Styles					
		Authoritarian	Democratic	Hands Off	Flexible
C h a r a c t e r i s t i c s		Leader sets specific time frames for project completion. Lays out all work expectations. Gives orders. No employee input. Control always with leader.	Seeks employee input. Encourages employees to participate in decision-making. Employees sometimes on their own. Employees sometimes work with input from leader.	Assumes employees can do the job alone. Assumes skills are there. Provides no direction or feedback. Feels employees will be motivated if trusted and left alone.	Observes motivation and skill level of employees. Assigns tasks according to ability. Provides level of supervision adjusted to individual employee and task. Uses style appropriate to situation and/or employee.
T y p e o f	**L e a d e r**	Unsure of others. Often insecure. Inflexible. Gives appearance of self-confidence.	Self-confident. Secure. Believes in staff capabilities. Believes in collective decision-making	Laid back or lazy? Feels employees must be responsible for their own work. Job satisfaction dominent.	Same as democratic except believes employee skills, nature of situation will determine style.
A d v a n t a g e		Works well when you need to send the troops charging over the hill, or in a short-term emergency.	Encourages participation. Increases self-satisfaction and commitment to team goals.	Great for self-starting, creative types who need no direction and have time for mistakes.	Allows Supervisor to use style appropriate to situation and employee. Very effective in developing employees.
D i s a d v a n t a g e		Demeaning to most employees. Ineffective with bright, creative people. Does not develop staff.	Some people don't want to make decisions. Some situations are not suited to collective decisions.	Bad for everyone else.	Requires Supervisors willing to learn and work smarter.

your observation of their behaviors, motivation and skills. Table 3 – 1 shows you a comparison of the four supervisory leadership styles. Study it, then do some practice work. Then practice some more. Then practice even more. That's what makes for really good supervisory leadership.

CONCLUSION

In this chapter we have seen that the Supervisor is clearly an important part of the management team. We looked at questions of how Supervisors exercise leadership in the workplace, exploring the range of behavior from authoritarian to democratic. We then looked at how this works in terms of a supervisory leadership style, defining four different styles: authoritarian, democratic, hands-off and flexible. After reviewing what motivates employees, it became apparent that if you are to work with a wide variety of employees who have different needs and skills, you will need to learn a flexible supervisory leadership style. That will take practice.

Chapter 4

Communication: A Two-Way Street

VERBAL

NONVERBAL

WRITTEN

TO: MAINTENANCE SECTI
From: Joe Smith
Supervisor
Date
Subject

Figure 4 – 1. Three Types of Communication

OVERVIEW

In this chapter we will be dealing with six key aspects of effective communication between Supervisors and their employees. These are:

- Looking at current skills
- The three kinds of communication
- Active listening
- Confronting problems through effective communication
- Dealing one-on-one with those you supervise
- Communicating with groups

Taken together, these provide a thorough review of the critical communication skills needed by Supervisors.

LOOKING AT CURRENT SKILLS

More than anything else, being a Supervisor means working with people. To work with people, you must be able to communicate. Therefore, being an effective Supervisor means being an effective communicator. If you have ever seen (or worked with) a Supervisor who didn't give instructions, or who mumbled a few words to you that you didn't understand, you learned a little about ineffective communication. In this chapter, we will be dealing with the skills needed for effective communication with your employees.

There are three types of communication: verbal (talking), nonverbal (body language) and written (Figure 4 – 1). We will discuss each in detail in this chapter.

Clear and precise communication leaves the impression that you are knowledgeable. Vague communication leaves the impression that you don't know what you're doing. If you want your team to respect you as a Supervisor, the choice is obvious.

You may think, *I don't have enough time in the day to worry about 'how' I communicate.* If that's the case, then ask yourself if you have time *not* to think about how you've communicated. For example, have assignments gone wrong for you because of poor communication? Have you seen it happen when others are supervising or managing you? If you've noticed this, you must pay attention to how you communicate.

Don't assume good communication happens naturally. It doesn't. Good communication is an acquired skill, and good communication skills are a result of thinking and practicing (Figure 4 – 2). Try to remember one or two work situations when you felt you had successfully communicated with others. These could be verbal, nonverbal (body language, gestures, eye contact) or written communications. For example, let's say you assigned a task, and it was completed exactly as you instructed and intended it to be done. That's pretty good communication. Write down one or two examples in Figure 4 – 3.

Now think of one or two work situations when you didn't successfully communicate either verbally, nonverbally or through written communication. As an example, perhaps you assigned a task that was done properly but not in the time frame you expected because you didn't make the deadline clear. Write down one or two examples of these problem situations in Figure 4 – 4.

Look over these examples and ask yourself:

- How could I have improved my instructions in the cases where things didn't work out?
- Was the problem mine or was it with the employees?
- Did I anticipate problems and communicate accordingly?
- Did I listen to what they were saying?
- Did I know what they could or couldn't do?

Listening to Feedback

When you are a Supervisor, you must keep your ear carefully tuned to feedback from employees and those above you. What kind of feedback have you received from your subordinates, peers or boss?

Do people often have to ask you to speak more loudly or clearly during a conversation? Then you have a communication problem, and maybe a hearing problem as well.

Has someone had to ask, *Are you listening to me?* Has someone called and said, *I'm a little confused by this memo you sent me.* These are examples of feedback that indicates there is a communication problem. You should listen to this feedback and act upon it.

You should seek out feedback. Ask for feedback from a trusted colleague, boss or friend who will give you honest, constructive input regarding your communication skills. It can hurt a little to receive some criticism, but it's the only way you have to eliminate communication problems.

You can also ask yourself what things you're doing that you really don't intend to do. Pretend you're a Martian looking down at yourself. What nonverbal clues are you sending to those you supervise? Do you give instructions like a drill sergeant? Do you walk around with a frown? Do you stand there with your arms crossed and folded in front of you? These can all be taken as signals of anger. Maybe you don't intend to convey anger, but that is how the nonverbal message is received. In today's very diverse workforce, the Supervisor must be conscious of the signals intended and the signals received. We will deal with the diverse workforce in detail in Chapter 6, Supervising a Diverse Workforce.

Try having your written correspondence read by a friend,

Figure 4 – 2. The "Art" of Communication

Good communication doesn't happen naturally.

Good communication is an acquired skill.

Good communication skills come from thinking and practicing.

Figure 4 – 3. When Did Your Communication Succeed?

EXAMPLE 1: _____
_____
_____
_____
_____
_____
_____

EXAMPLE 2: _____
_____
_____
_____
_____
_____
_____

Figure 4 – 4. When Did Your Communication Fail?

EXAMPLE 1: _____
_____
_____
_____
_____
_____

EXAMPLE 2: _____
_____
_____
_____
_____
_____

such as your spouse, a parent or older child, and ask for interpretation of the meaning to see if the message that is received is the same one that is intended.

Finally, think about those people you know who appear to have good communication skills. How do they do it? Think about the techniques they use.

THREE KINDS OF COMMUNICATION: VERBAL, NONVERBAL AND WRITTEN

Good communication involves three aspects: talking, using body language and writing. To be an effective communicator, you must master all three of these ways of communicating.

Verbal Communications

We may all know how to talk, but how well we can communicate with others in the workplace is a different matter. There are some fairly simple steps you can take to improve your verbal communication as a Supervisor (Figure 4 – 5).

Plan your messages. You should plan what you intend to communicate. Off-the-cuff instructions seldom produce the intended results. Your employees are not mind readers, so don't expect them to be good at reading your intentions from a few casual remarks. Decide what you want done and convey the message clearly. To do that, you have to plan your message. Ask yourself, *What is the objective of this communication?* Then clarify your ideas so that the point of the communication is not clouded by miscellaneous chit-chat. This doesn't mean being abrupt or unfriendly. It does mean thinking about your objectives.

Consider the receiver. Just who is receiving the message you're sending? Ask the following questions about the receiver:

- What type of communication does he/she respond to best? (Joe always responds better if you talk to him alone. Ann likes to hear all of the details of an assignment.)
- When and where will the receiver be able to listen to you without any undue distraction? (When machines are pounding away on the shop floor, it's usually helpful, if possible, to get away to a quiet place before giving a complex instruction.)
- Does the receiver have any biases or strong feelings about the message you plan to communicate? (Jim has always preferred to keep a large supply of parts on hand, and you have to communicate to him that this will no longer be possible.)
- How can you communicate so as to help the receiver keep an open mind regarding your message? *Jim, I know you like to keep a good supply of parts on hand, but we're going into a tight inventory control process, and I'd really appreciate your not storing things up for a while. Let's give it a try this new way, and we'll both check it closely to see how this new system works.*

Consider how you sound. Have you ever listened to your-

Figure 4 – 5. Effective Verbal Communications

PLAN YOUR MESSAGES

Know what you intend to communicate.
Don't expect people to be mind readers.
Convey the message clearly.

CONSIDER THE RECEIVER

Find out what communication gets the best response.
Plan when and where the message will be received.
Understand the biases or strong feelings involved.
Communicate so as to encourage an open mind.

CONSIDER HOW YOU SOUND

Consider the tone and volume of your voice.
Work to clarify your language.
Select your words.

CONSIDER YOUR RATE OF SPEECH

Too fast?
Too slow?

HOW'S YOUR LANGUAGE?

Are you using the right words?

KEEP IT SHORT AND SIMPLE

ASK QUESTIONS TO ASSURE FEEDBACK

self on a tape recorder and noticed that your voice sounds different from what you hear in your head? In a similar way, what other people hear when you speak is different from what's in your head. They must interpret your tone and attitude without all the thoughts in your mind. There are some key things you can do to improve the sound of your communications.

- *Tone and volume of voice.* Is your tone of voice communicating anger, impatience, lack of interest? Does your tone change, or is it monotone? Do you speak loudly enough for the receiver to hear you? Do you speak so loudly that everyone in the room can hear you?
- *Rate of speech.* A very fast pace, or a very slow pace of speaking can be distracting. Fast speaking causes people to miss words. Speaking too slowly causes boredom and inattention to what you are saying. Listen to those around you. If they sound slower or faster than you, you will improve communication if you try to follow their pace.
- *Clarity of language.* Articulation and choice of words is important. Articulation means getting the whole word out without swallowing huge parts of it, like *Ya'll come!* or *Whadaya mean?* or *Dintchya unnerstan me?* To articulate, you must speak clearly so the receiver can understand you. Avoid leaving off the end of words, such as *goin'* instead of *going;* avoid using dialect words, such as *ax* instead of *ask;* and avoid using *um* or *uh.*
- *Word choice.* Choose the most appropriate word for the message and the receiver. Avoid words with double meanings or words or phrases that can offend or be misinterpreted by the receiver. A simple rule of thumb when giving instructions: keep it short and simple. A lot of Supervisors call this the KISS formula: Keep It Short and Simple.
- *Jargon.* Using jargon is only effective if all the people you are dealing with understand it. If they don't (which is probably the case), they feel stupid, or you sound stupid. Computer buffs who talk in terms of bits, bytes, RAM and k's can understand each other. The person trying to learn how to use a computer doesn't have a prayer of understanding. If everyone in your department understands that *jadpics* stands for the Johnson Automatic Digital Process Instrument Control System, use it. But don't try to use the term when communicating to those outside the department.

Ask questions. Do you ask questions to clarify understanding when you have issued some instructions or assignments? If you don't, you're depriving yourself of the feedback necessary to make sure things are done right. That means you're setting yourself up for a fall. Start asking those questions to check on the delivery of your communication to the receiver, to see if he or she has interpreted your message as you intended.

Ask open-ended questions to elicit a response. For example, ask *What do you think we can do to improve on this approach?* instead of *Do you understand?* An open-ended question has to have an answer other than Yes or No. Asking open-ended questions means you'll get the feedback you're looking for.

Nonverbal Communication

Nonverbal communication covers a lot of behavior. For example, think about the Supervisor who walks by a milling machine, sees that a worker is not wearing safety glasses, but turns his back and walks away. He has communicated with his eyes and his body. The communication is clear: *I don't care if you violate the safety rules.* That may not have been his intended message, but it certainly was the received message. And all that happened without a word being spoken! Approximately 75–80 percent of what you communicate is done through nonverbal means. That seems like a startling amount, but just look at what constitutes nonverbal communication (Figure 4 – 6).

Eye messages. Eye contact expresses interest in what you are communicating and your interest in the receiver(s). Lack of eye contact indicates that you are not interested in what you are communicating or the receiver. Think about it. What is your reaction to people who won't look you in the eye, or who always shift their eyes away from you when they give an answer. They're sending messages. How do you interpret these messages?

Posture messages. What does your posture say? Turning your back to someone during a conversation will likely send a message of dislike, disrespect or lack of importance of the subject. Good posture, on the other hand, will generally send a message of self-confidence and authority. A slouching, laid-back posture sends a slouching, laid-back message.

Hand messages. Hand gestures send a lot of messages. We all know some pretty clear and pretty crude examples of this. But there are other everyday hand gestures that can send messages just as negative, even when unintended. When a Supervisor points a finger at an employee it will almost always be interpreted as accusatory or reprimanding or condescending. If you've ever had it done to you, you understand this. So why do it to others? Holding your hand up like a traffic officer indicating Stop can cut off all lines of open communication, even if you were only looking for a pause. On the other hand, a good firm handshake demonstrates self-confidence and a sincere message of *It's nice to meet you* or *Thank you for doing a good job.*

Body messages. Body language is a popular phrase that refers to all the nonverbal messages your body sends out even when you don't know it. We discussed the anger example earlier: frowning face, tightly crossed arms, spread-leg stance. In contrast to this, uncrossed arms and legs often communicate that you are open to the ideas being communicated as well as open to the communicator.

Sometimes even the distance between the communicating individuals sends a message. As you move across the United States, you find that people in different areas of the country stand at different distances from each other when they are talking. Easterners stand 12–18 inches apart (a closeness that makes Westerners very uncomfortable). Midwesterners stand 20–30 inches apart, Westerners 36–40 inches, and West Coast residents

Figure 4 – 6. Nonverbal Messages

- Eye messages
- Posture messages
- Hand messages
- Body messages
- Dress messages
- Consistency between messages

stand 20–24 inches apart. That is what is called the social distance. When you move too far into or away from the regional customary social distance space, people get uncomfortable.

Supervisors who get transferred from one area of the country to another, or who deal with people from different areas of the United States or even different countries should be aware of these regional differences in speaking distance and the body messages being sent. Ethnic and cultural differences complicate this even more: Mediterranean people tend to stand closer together and touch each other; Germans stand apart with little direct contact; Japanese stand quite far apart and are very formal, and so on. All of these, of course, are generalizations and there will be exceptions to each.

Dress messages. How you dress for your job does make a difference in most supervisory situations. Is the way you dress for work consistent with your supervisory position? In almost every work situation, Supervisors do dress somewhat differently from those they supervise. We've been told that the Japanese don't do things this way, but that's because our cultural eyes are not trained for seeing their distinctions in dress. Japanese Supervisors may appear to be wearing the same coverall uniform worn by the workers, but a careful eye will note subtle distinctions in rank, such as stripes or marks similar to those on military uniforms or simply the quality of the materials and tailoring. The same distinctions were found in Mao's supposedly classless military forces. The distinctions are there, and every worker knows it.

For Supervisors, these distinctions cover a vast range of dress, depending on the industry, nature of work, local workplace, and so on. In some places, male and female Supervisors must wear suits. In other places, shirts and ties, coveralls, lab smocks or certain types of caps form distinctions between Supervisors and those they supervise. Like it or not, if you don't make these dress distinctions in the American workplace, those above you and those below you will not show you the respect a Supervisor needs to be effective.

Consistency between verbal and nonverbal messages. Is your message consistent with your action? Do you send contradictory messages to your team? Do you ask for suggestions but never act on them? Do you follow through on verbal commitments? Does your body say one thing while your words say another? Let's look at an example.

CASE STUDY: MIXED MESSAGES

Jim, a maintenance mechanic, had 25 years of service with the company. A luncheon, given in honor of the occasion, was attended by higher management. Jim was congratulated and complimented on his years of faithful service. His loyalty and value to the company was dwelt upon in speeches by these higher management persons, all of whom patted him on the back, shook his hand, looked him in the eye and otherwise oozed sincerity. Jim enjoyed the party and was pleased with all the attention given him.

Ten days later, Jim, obviously angry and upset, burst into his Foreman's office and threw his 25-year pin on the desk. Words tumbled out so fast that it was hard to understand them. It seemed that Mr. Johnson, the General Manager, who had made one of the speeches about Jim's "loyalty and value to the company" had come through the plant. Even though Mr. Johnson practically stumbled over Jim and looked straight at him, or as Jim said, *right through me,* he didn't greet Jim or give any sign that he even knew him. Jim put it directly: *What hypocrites all these top management people are! I just wish Johnson was here so I could tell him what to do with his lousy pin. And that includes, you, too,* Jim concluded. With those words, Jim stormed out of the office.

As the Foreman, you're stuck with the whole problem. Those at the top obviously sent mixed messages to Jim. They pretended sincerity with their words and their body language. But they didn't even take the trouble to look at Jim as an individual. They made a mistake and you heard about it from Jim. People do read body messages and listen to verbal messages. And they hate inconsistency.

Written Communication

Written communication is probably the most overused form of communication in today's workplace. Between computers and copy machines, our ability to generate tons of written communication has been on a steep upward increase. There was some hope that computer electronic message and mail systems could reduce some of this. It appears that while these systems seem to have reduced the amount of paper being sent, they have not reduced the amount of on-screen messages.

So, if there is a first rule in written communication for Supervisors, it is only use it when absolutely necessary. It's a much better rule of thumb to deliver a message in person so that you can allow for input, feedback and personalization.

Advantages. There are, of course, some advantages to written communications in a supervisory situation. We suggest these generally come as a follow-up to verbal communication. In that context, the advantage of written communication is that it:

- reiterates something said verbally,
- provides a record of the message delivered,
- may be taken more seriously if put in writing.

Potential pitfalls. There are three major pitfalls to written communication. Written communication does not always offer the opportunity to check if the intended message was the message received. Also, written messages are usually impersonal and don't build rapport as verbal communication can. Finally, written communications tend to freeze flexibility. You wrote it, and now you have to live with it.

Guidelines for effective written communication. These

guidelines are the same as for verbal communication, since both, obviously, depend on words (Figure 4 – 7).

- Plan your correspondence with an objective in mind.
- Keep the receiver in mind.
- Choose words carefully to avoid jargon, slang or words that could be misinterpreted. (Somehow, when we write, we always make things more complex than when we speak.)
- Keep sentences and paragraphs short. Get to the point, and keep it brief.
- KISS

ACTIVE LISTENING

Active listening is a concept used in human resource management to focus on the need to really work at listening to people. Active listening is one of the most useful skills a Supervisor can gain for dealing with people. It doesn't take much study, just some practice, and it really works!

Listening

Since we all listen, we think we know all about it. You just sit back and let the ears work. But listening is not a passive activity. It takes effort, concentration and practice. In everyday life and in your job as a Supervisor, how well you listen sends a message. How well you listen also influences how much you hear and how effective you will be. Effective listening means "actively" listening. It takes practice. If you become an active listener, you will have acquired one of the most useful supervisory skills available to you (Figure 4 – 8).

Tips for Becoming an Active Listener

There are many different ways you can improve your active listening capacity. The following tips will help the Supervisor who is actively listening to a problem presented by an employee.

- **Have a positive attitude.** Wanting to listen is half the battle.
- **Practice.** We often aren't accustomed to really listening to others. Try it out with friends or your spouse, in all sorts of places outside of work. Even the Sunday sermon may take on new meaning if you actively listen to it.
- **Avoid distractions.** Turn off the radio. Find a quiet place to converse if the plant floor is noisy. Have someone else answer your phone. Close your door. Clear off your desk. Avoid too much notetaking so you can concentrate on what's being said. Don't doodle.
- **Schedule the discussion.** When you schedule time for important conversations, you can plan ahead to avoid distractions or interruptions.
- **Postpone if distracted.** Postpone conversations if you are distracted and can't change that situation at the moment. Maybe you're angry, tense or nervous, and you need

Figure 4 – 7. Guidelines for Effective Written Communication

- Plan your correspondence with an objective in mind.
- Keep the receiver in mind.
- Choose words carefully to avoid jargon or slang.
- Keep sentences and paragraphs short.
- KISS

Figure 4 – 8. Active Listening

Confucius says:

"Man was given one mouth and two ears so that he can listen twice as much as he speaks."

to listen at a later time when you're not preoccupied. But do set a time.

- **Be aware of your biases.** Don't jump to conclusions. Concentrate on what's being said, not how it's being said or by whom. Try to be open to new ideas and concepts.
- **Be empathetic.** Try to take the speaker's point of view or frame of reference when you are listening.
- **Listen to the whole message.** Really wait for the whole message to be put forth, and don't filter out facts along the way. You may miss the whole concept by selectively listening to only certain facts.
- **Watch for nonverbal clues.** What message are they sending?
- **Avoid quick answers.** Don't formulate an answer or response before the speaker has finished. S. I. Hayakawa, a noted linguist, stated:

  Listening doesn't mean simply maintaining a polite silence while you wait for a conversational opening. Nor does it mean noting the flaws in the other person's argument so you can show him up.

- **If you can't hear, ask.** Don't let people mumble along and miss half the message they are trying to give you. Politely ask the person to speak up.
- **Listen to and be sensitive of feelings, emotions and unspoken needs or concerns.** Respond with concern, but carefully, remembering you are not a psychiatrist.
- **Be patient.** Patience and holding your temper is absolutely critical to your listening success.
- **Ask clarifying questions,** but don't play Perry Mason and start a cross-examination. Always keep in mind that your primary purpose is to listen.
- **Send interest messages**. Use nonverbal clues that show active listening, such as eye contact, open body positioning and posture.
- **Sum up.** Sum up and paraphrase what has been said to insure that you understood and interpreted correctly.

If you practice these techniques, you will find that active listening not only leads to clearer communications but also sends a positive and motivating message to your employees. It fosters open communication between yourself and those you supervise. It will even improve communication with your boss. You will be a far better Supervisor for it (Figure 4–9).

CONFRONTING PROBLEMS THROUGH EFFECTIVE COMMUNICATION

Clear, direct communication is the best way to confront workplace difficulties. As stated earlier, your employees can't read your mind. If there is a problem, communication is usually the means of resolution.

Clear, direct communication is also the best way to avoid workplace problems or conflict in the first place. Communication

Figure 4 – 9. Tips for Becoming an Active Listener

- Want to listen
- Practice
- Avoid distractions
- Schedule discussions
- Postpone, if distracted
- Know and control your biases
- Be empathetic
- Get the whole message
- Watch nonverbal clues
- Avoid quick answers
- If you can't hear, ask
- Listen with concern to feelings, emotions and unspoken needs; respond carefully
- Ask clarifying questions only as absolutely necessary
- Be patient
- Send interest messages
- Sum up
- Practice, practice, practice

requires at least two people to be effective. If you keep things to yourself, you are obviously not communicating. Yet, some people forget that simple fact.

CASE STUDY: COMMUNICATION

Jerry Brown, the Supervisor, was assigning his plant maintenance employees specific tasks for the morning. He assigned Oliver Clawson the job of troubleshooting a defective electric motor in Building B and suggested that Ollie take with him the rasp needed to smooth a rough commutator. As Ollie was heading out the door, Jerry noticed that the rasp was left behind on the bench. Before he could call out to Ollie, something else took his attention away for the moment. By that time, Ollie was beyond hailing distance, and Jerry said to himself, Well, I guess he didn't hear me, or didn't listen. Anyway, he may not even need the rasp, and if he does, it's only a short walk back to the tool room to get it. Besides, he knows he should have all of his tools with him.

However, wishful thinking aside, this was a job for which the electrician did need the rasp. As it happens, when Ollie returned for it, he ran down the stairs without holding the handrail, slipped, fell and broke his ankle!

The moral of this story is not that an employee will break his ankle every time you forget to remind him of something or repeat something you've already told him! The injury simply illustrates that accidents are often related to haste and distractions. Ollie was distracted because he realized he had forgotten the tool, needed it and was rushing to get it and return to the job.

That sequence of events leading to a broken ankle and an unfinished job could have been avoided in three ways.

- First, the employee should have taken more care with his tools and job preparation.
- Second, the Supervisor should have acted at the moment when he realized that the employee had either not understood his suggestion or forgotten it. He needed to catch him right then, confront the problem when it was first observed. Don't engage in wishful thinking.
- Third, the employee should have returned without haste, observing safety rules.

The only one of these three preventive actions under the Supervisor's control is the second. When you see something wrong, act. Confront the problem and use clear and precise instructions to convey what must be done. If this theme sounds familiar, it will become even more so as we go through *Out In Front*. One of the great weaknesses found in Supervisors (and managers) is their inability to confront inappropriate behavior in the workplace, whether in safety, job preparation or performance. Problems seldom disappear, but overlooking or neglecting them is a very effective way to make sure they get worse (Figure 4 – 10).

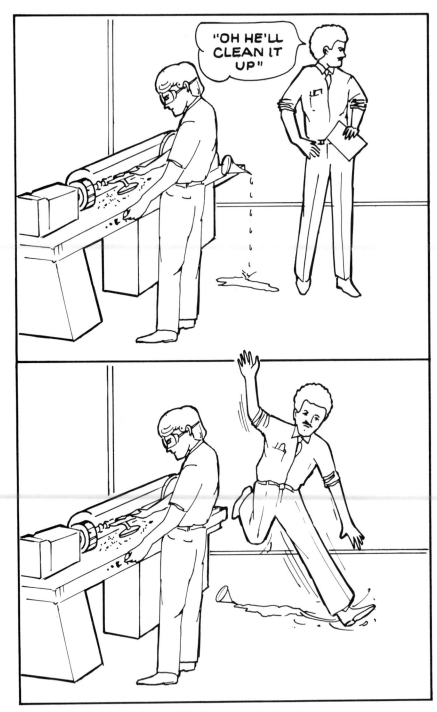

Figure 4 – 10. Unconfronted Problems Only Get Worse

ONE-ON-ONE WITH THOSE YOU SUPERVISE

We have made a major point of the fact that you must deal one-on-one with those you supervise. Sometimes, it seems so obvious there is little to say about it. Yet, it is in the failure to do this that many Supervisors flounder. Therefore, let us restate the obvious.

PEOPLE NEED TO BE TALKED WITH AS INDIVIDUALS.

You can talk with your employees as a team, you can talk with them in smaller groups. You can have a lot of good-old-boy (girl) stuff going on. But:

PEOPLE NEED TO BE TALKED WITH AS INDIVIDUALS.

Sometimes it's hard to talk to some people, but you have to try. You must encourage open lines of communication with your employees. They need to know they can come to you with a problem or concern. You must send consistent messages and listen actively to develop their trust in you and the fact that you really are interested in them. The bottom line is, if you are to have an effective team, you must reserve time for one-on-one communication with your employees.

PEOPLE NEED TO BE TALKED WITH AS INDIVIDUALS.

COMMUNICATING WITH GROUPS

Supervisors are frequently called upon to deal with groups, from a small team to a presentation to a group of teams and even to an occasional presentation to management. Communicating with groups is different from communicating with friends and individuals. The principles of good communication, however, continue to apply.

Calling a Meeting

RULE 1. Don't call a meeting without a purpose. Respect your staff by not wasting their time in a meeting with no purpose.

RULE 2. Don't call a meeting if the problem can be handled with one or two people and doesn't involve the whole group.

Both of these rules can be followed by asking yourself: What is my purpose, and is this purpose appropriate for a group setting?

Leading a Meeting

Once the purpose of the meeting has been established, only invite staff that need to be there. Now go through the following steps:

- Consider the number of attendees and effectiveness of the size of the group. Keep in mind the purpose of the meeting while doing this.
- Prepare and distribute an agenda. Follow the agenda and keep to the time frames established. It is usually helpful if everyone knows the purpose in advance and, ideally, has a copy of the agenda in advance. This allows everyone to prepare for the meeting.
- Open the meeting by stating the purpose. This provides focus and direction. (This holds for one-on-one interaction also.)
- If the purpose of the meeting is merely to disseminate

Figure 4 – 11. Checklist for Preparing for a Meeting

_____ Do I have a specific purpose?

_____ Who needs to be at the meeting?

_____ Is the size of the group manageable?

_____ Are meeting room arrangements made?

_____ Have I prepared and distributed an agenda?

_____ Do I have a clear statement of purpose for opening the meeting?

_____ If the purpose of the meeting is to disseminate information, have I made that clear?

_____ If the meeting is to solve a problem, do I have a clear statement of the problem and background information?

_____ Do I have open-ended questions prepared to encourage participation and aid my active listening?

information, make that clear so the meeting doesn't become a debate.

- If the meeting is to solve a problem, state the problem and any background information.
- Spend most of your time directing the conversation by actively listening, paraphrasing to clarify understanding and asking open-ended questions to gain information and suggestions from participants.
- Try to sum up issues and suggestions.
- Prepare minutes of the meeting, outline decisions reached, responsibility assigned and expected time frames.
- Distribute minutes as soon after the meeting as possible (Figure 4 – 11).

Difficult Group Members

There are always those who make it a little hard to run meetings. Some are earnest, they just want a lot more detail and background. Some are people who like to show off their knowledge. Some really want to please you and keep asking friendly questions that nonetheless divert you. Some are just plain troublemakers. The most common ways of handling these problems in a meeting are:

- Restate the purpose of the meeting or topic if people start to get off the track.
- Ask for input from participants who haven't had a chance to offer their input either because they are shy or someone else is trying to dominate.
- No matter what, don't belittle a participant in front of the group.
- Remember that it is your meeting.

Informal Discussions

Informal discussions between the Supervisor and her/his crew is another form of "meeting." Informal discussions help build rapport with your staff and encourage open communication. They are a very effective supervisory tool.

Informality can sometimes be a bit frightening to the new Supervisor (particularly when he/she has been brought in from outside), since the team members know each other a lot better than you know them. Nonetheless, informal discussions can be handled so that they do not lessen your authority or position in any way. Basically, you need to apply the same approach as recommended for dealing with difficult group members, but you do it informally since you are not dealing with difficult employees. Thus, you won't say something like *Let's all remember that our purpose in this meeting is to deal with the equipment breakdowns in Section 2,* but would instead mention several times how Section 2 certainly seems to be giving us problems.

CONCLUSION

This chapter has explored six key aspects of effective communication between Supervisors and their employees: (1) looking at your current skills, (2) dealing with the three kinds of communication, (3) active listening, (4) confronting problems through communication, (5) dealing one-on-one with those you supervise and (6) communicating with groups. Each of these key aspects of effective Supervisor/employee communication skills will affect your performance.

When you look at your current communication skills do you: Think about the kind of instructions you give? Honestly look at your failures and successes? Listen to feedback from others? Think about the clarity of your writing? Try to model yourself after someone you know who is a good communicator?

Have you learned the importance of and problems with the three kinds of communication: verbal, nonverbal and written? Do you know the all-important KISS principle? Do you understand active listening and how you can apply it in your work? Are you prepared to confront problems, and do you understand why you can't simply ignore them? Do you know how to deal one-on-one with those you supervise and why it's a good idea to do it? Finally, do you understand how to effectively communicate with groups? These are the things we have covered, and these are the tools you will need to be a good communicator and Supervisor.

Chapter 5

Talking to the Troops, or Talking with the Team

OVERVIEW

In this chapter we will be exploring five key techniques used by good Supervisors in human resource management. By techniques, we mean a part of the job that involves a How-to-Supervise matter, rather than a description of parts of the Supervisor's job. These five techniques all involve the combination of communication and some of the important roles the Supervisor must carry out. Most importantly, they show how the Supervisor must function when dealing with employees and employee problems. The five techniques are:

1. Creating employee participation
2. Delegating responsibility
3. Team building
4. Handling complaints: gripes and grievances
5. Applying the rules

In each of these techniques, a Supervisor has to make a choice between talking to the troops or talking with the team. That choice reflects both supervisory leadership style and, putting it simply, good management. At the heart of it all is the first technique: creating employee participation.

CREATING EMPLOYEE PARTICIPATION

Employee participation is a very necessary part of good supervision and good management. It has been a key principle of modern American management for over 30 years. Employee participation is, in fact, the single most important technique available to the Supervisor. The popularized current version of Japanese management style, with its heavy emphasis on employee participation, was actually drawn from American management theory and practice.

Employee Participation Motivates

Encouraging employee participation helps motivate your work unit by recognizing that the employee is worthwhile and has worthwhile ideas. This sense of recognition of the individual's worth in the workplace is part of what we try to achieve in one-on-one communication. When the individual understands that you really do value her/him and the work produced, his/her sense of involvement and commitment to the work, to you and the organization, will increase tremendously (Figure 5 – 1).

Employee Participation Increases Productivity

When the Supervisor uses participation as a motivating tool, it increases an employee's sense of self-esteem and recognition and is likely to increase productivity and produce a higher quality of work. How do you, the Supervisor, achieve this? By building in a set of Do's for yourself that really implement the idea of recognition. These are:

- Do reward employees' suggestions or initiatives intended to improve productivity or quality.
- Do praise employees for showing initiative and interest in improving operations, projects and jobs.
- Do set up formal and informal systems for new ideas, such as suggestion boxes, special awards, times for discussions, incentives for dollar-saving proposals.
- Do encourage continued effort even when initial suggestions don't work!

Employee Participation Creates Ownership

When employees have a true sense of involvement with their work and decisions related to it, they develop a strong sense of ownership of the project. That ownership is most often expressed as trust and loyalty to the company and the Supervisor. This is a very important part of what makes a good employee. And good employees reinforce and support a good Supervisor. Funny, isn't it, how the good Supervisor encourages and supports employees who, in turn, encourage and support the good Supervisor?

How to Promote Employee Participation

The following section shows you examples of things that can be done to promote participation (Figure 5 – 2). The comments and suggestions included here are drawn from extensive experience.

- Always talk *with* employees instead of to or at them. This way, you allow for participation, and they'll know you want their suggestions.
- Always be assertive rather than aggressive. Aggressive Supervisors talk at their employees, are poor listeners and make decisions for their employees instead of with their employees. This shows they're really tough guys or gals. Assertive Supervisors encourage and build open, two-way communication for ideas and suggestions. At the same time, they know their goals and objectives and are able to get them across to their staff.
- Always ask open-ended questions. These help you to know what employees think. Yes or no answers to your questions won't be enough.
- Use open-ended phrases at the beginning of questions,

Figure 5 – 1. Employee Participation

WHEN IT COMES TO GOOD SUPERVISION,

EMPLOYEE PARTICIPATION IS NOT A CHOICE. . .

IT'S A REQUIREMENT.

Figure 5 – 2. Tips on How to Promote Employee Participation

Always talk with your employees.

Always be assertive rather than aggressive.

Always ask open-ended questions.

Use open-ended phrases to preface questions.

Repeat the employee's key words, phrases or ideas that are on track.

Summarize conversations to make sure you've understood or been understood.

Ask for help in solving problems.

Always use good employee suggestions or ideas, giving the employee credit as the originator.

Praise employees for good ideas.

Always let employees know the result of their participation.

such as *How about...?* or *What do you think?* Questions like this encourage a lot of communication.

- Repeat key words, phrases or ideas when an employee is on-track since this will encourage him/her.
- Summarize conversations to make sure you've understood what employees have suggested or that they've understood your directions.
- Ask for help in solving problems.
- Always pick up on employee suggestions or ideas that you like, being sure to let your boss know they are your employee's new ideas.
- If an employee makes a good suggestion, praise him/her for it.
- Always let employees know what the result of their participation is. Let them know why suggestions weren't able to be implemented, and certainly let them know when they have been.

Time, Place and Attitude

We do not mean to imply that everything you do should be run through a meeting with employee participation. To the contrary. Much of what a Supervisor must do each day, such as checking schedules, observing performance, ordering materials, does not need and would not be improved by employee participation. But the people side of things, such as getting everyone to pull together, having a clear common purpose, establishing cooperative relationships, is absolutely dependent on effective employee participation. It is a major supervisory technique.

At the same time, you must recognize that while encouraging and supporting employee participation is a vital supervisory tool, not all employees want to or will become involved in active participation. Some employees are simply reluctant to participate. Others believe deeply that their job is to do what you tell them to do, nothing more and nothing less. Still others seem to be waiting around to be fired.

Employee participation, under these circumstances, should be encouraged, but it should not be considered mandatory for employees. It should be considered a mandatory tool you must use as a Supervisor, but only when appropriate.

Thus, you must be selective about the use of the technique, just as you must be selective in the way you treat different employees. Group decisionmaking about when to order routine supplies would probably be a waste of everyone's time. Group discussion of ways to cut costs in maintaining an inventory, on the other hand, could probably benefit significantly from the input of those who use that inventory on a daily basis, your employees. That's effective employee participation.

Finally, one of the most frequent uses of employee participation is delegating responsibilities, new assignments or key projects. That's what we will discuss in the next section.

DELEGATING RESPONSIBILITY

The second technique of supervision, successful delegation of responsibility, is dependent upon two main factors: the willingness of the employee to accept responsibility for the assigned work and the Supervisor's willingness to give up some portion of control of the work to the employee (Figure 5 – 3). We will explore these two factors throughout this section.

The first question we must address is Why should the Supervisor delegate responsibility?

Why Delegate?

Why should Supervisors delegate part of their responsibility to employees? First, it encourages employees to feel that sense of self-worth, involvement and ownership of the work. Second, it helps improve productivity and quality by bringing more skills and ideas together to address common problems. Third, it allows you to work smarter by having others pick up on some of your work, thus making it possible for you to concentrate on other parts. Fourth, you're not Superman or Superwoman and could probably use a little help now and then (Figure 5 – 4).

Remember, as discussed in Chapter 3, that although a Supervisor delegates responsibilities to an employee, the Supervisor does not, in that way, become absolved of responsibility. Ultimately, a Supervisor is always responsible for anything delegated to others. In that sense, the delegated responsibility is shared.

What Should Be Delegated?

Delegation of responsibility usually means the Supervisor passes on some portion, if not all, of an assignment or task to an employee or group of employees. What to delegate is answered by asking yourself the following questions.

Is this an assignment you can delegate or does your boss specifically want you to complete the assignment? Sometimes the boss is very clear about this, most times not. You can learn to judge this, and also to coach the reluctant boss into a more participative style by pointing to the successes of your team's handling of delegated tasks.

Can the whole assignment be delegated or only pieces of it?

Figure 5 – 3. Delegation of Responsibility

Successful delegation of responsibility depends on:

1. The employee's willingness to acccept responsibility.

2. The Supervisor's willingness to give up some portion of control to employees.

Figure 5 – 4. Delegating Responsibility: Why?

It encourages employee self-worth, involvement and ownership of the work.

It improves productivity and quality.

It allows you to work smarter by having others pick up on some of your work.

You're no Superman or Superwoman and could use a little help now and then.

Figure 5 – 5. Delegating Responsibility: What?

Delegate some portion of an assignment if:

1. The job is one you're allowed to assign

2. The whole assignment can be delegated or only pieces of it

3. If you are willing and able to give up some control, then, as the saying goes, Try it, you'll like it.

Figure 5 – 6. Delegating Responsibility: To Whom?

Delegate to the lowest level possible to do the job.

Determine to whom to delegate by asking:

Who has the necessary skills or knowledge to complete the assignment?

Has anyone indicated an interest in this project?

Is this task an opportunity for the employee to develop present skills or to learn new ones?

Does the task fit specifically into someone's job description?

Can you spread it around?

Figure 5–7. Delegating Responsibility: When?

Delegate:

When there is time to delegate.

When it makes sense to do so.

Whenever possible.

Look at all of the parts of the assignment. There is a great likelihood that at least one of your employees can accept responsibility for parts of the assignment.

How much control are you willing and able to give up? This is usually a matter of your personality and style. As the saying goes, Try it, you'll like it (Figure 5 – 5).

To Whom Should You Delegate?

One rule of thumb is to delegate to the lowest level possible. That means people with the minimum skills necessary to perform the task involved. As a Supervisor, ask yourself if you are keeping assignments that could be handled by one of your staff. When deciding to whom to delegate, consider the following characteristics of your employee team:

- Who has the necessary skills or knowledge to complete the assignment? Even though they may not be as good as your skills and knowledge, are they sufficient to do the job?
- Has anyone indicated an interest in this project or task? If not, is there anyone you could encourage to get involved?
- Would the delegation of this task be an opportunity to develop someone's present skills or to teach new ones? This could really pay off for you and the employee in the long run.
- Does the task or assignment already fit specifically into someone's job description? If so, what is that employee's current workload, and would he/she have the time for this assignment?
- Can you spread it around? Don't always delegate to the same one or two favorites or trusted workhorses. Spread the work out to give everyone a chance to demonstrate abilities.

If you can't answer these questions positively, either the task shouldn't be delegated or you're not trying very hard (Figure 5 – 6).

When Should You Delegate?

The when of delegating usually has more to do with time than anything else. You should delegate when there is time to delegate and it makes sense to do so. In an emergency situation you will probably not have time to think about delegating, although some small parts could well be accomplished faster if they are delegated. Even in emergencies, try not to always do all of it yourself. If every day is an "emergency," then you should stop kidding yourself and start delegating. Delegate whenever possible and sensible (Figure 5 – 7).

You should especially try to delegate when the assignment of a project could be viewed as a reward. If a project clearly carries with it importance, the recipient of the delegation may be honored to be trusted with the assignment.

How Should You Delegate?

Delegation can be done verbally or in writing. You can delegate responsibility to an employee by requesting that something be done, or you can make a specific assignment. The first method implies consensus, you and the employee agree that the employee will carry out the task. That's almost always the best way. The second approach implies an absence of choice on the employee's part, even if you didn't mean it that way. Whatever you do, how you delegate depends on how you communicate. Most employees are likely to respond more affirmatively to a participatory process, that is, seeking cooperation on the request. *Can you help me with this?* goes a lot further than *Do it now!* However you do things, there are some other important steps involved in delegating.

- Clearly define the purpose of the delegation and the objective that is to be met. If the assignment is only a piece of a larger project, explain the whole picture so the employee has the opportunity to understand the importance and relevance of the assignment. This assists in creating employee ownership of the project. Such ownership usually means a much greater commitment to success.
- Define what is expected of the employee. How much latitude does he/she have in carrying out the assignment? Can he/she make decisions along the way or only do the limited steps you assigned?
- Explain whether the employee should check with you along the way or only check in when the task is complete.
- Be very clear about what the end result should be or will be. To say *Get the job done* is not sufficient. It leaves out how, when, in what shape and all of those other necessary detailed expectations you actually have.
- Set and communicate a clear time frame for task completion. The time frame can be a deadline date, or you may wish to establish checkpoints along the way.
- Identify who else is involved, if appropriate. For example, others may be working on the project with the employee, or other staff may need to be kept informed of the status of the project.
- Be very clear about what assistance or guidance you as the Supervisor will be providing. Make sure the employee has all of the resources necessary to succeed. Be sure to offer the employee reassurance and positive reinforcement throughout the assignment. Sometimes a *Hey, I read your notes yesterday on the project and am really pleased with the progress. Keep those notes coming!* is just what is needed to keep things moving.
- Check to be sure the employee understands what you have tried to communicate. Allow time for feedback or questions from the employee and watch for nonverbal clues that indicate understanding. Ask key questions or procedural questions to see if the employee understands the assignment.
- Follow up occasionally as the project progresses to see if the assignment is on track. Remember, however, if you have promised the individual complete authority over the project, checking up will create the impression that you don't have confidence in the employee's ability. This is

Figure 5 – 8. Delegating Responsibility: How?

Delegate verbally and/or in writing.

Make a request or make an assignment.

Clearly define the purpose and objective.

Define the expected results.

Clarify if the employee is to check with you along the way or only when the task is complete.

Set a clear time frame for task completion.

Explain who else is involved.

Be very clear about what assistance or guidance you will be providing.

Ask questions to ensure the employee understands what you have tried to communicate.

Allow time for feedback or questions.

Follow up as the project progresses.

discussed in more detail in the next section, because follow-up is so very important (Figure 5 – 8)!

Follow Up Your Delegation

The amount of follow-up a Supervisor must carry out is dependent on the level of authority and control you have allowed the employee. If you have agreed to delegate full responsibility, don't continually check up on the employee because this will undermine his/her efforts. If you have established specific checkpoints along the way, keep to this promise so a new or novice employee doesn't go astray. But remember:

- Even experienced employees may need special support when handling a new project.
- Positive feedback and reassurance is always important.
- Ensure that necessary supplies, tools and materials are available.
- Provide your support as Supervisor by serving as liaison with other departments to resolve any internal problems.
- Always follow up with the employee to let him/her know what has been done particularly well and what methods may need improvement or correction.
- If possible or appropriate, let the employee know the outcome of the work. Notice how these techniques apply in the following case study.

CASE STUDY: FOLLOWING UP DELEGATION

Upper management feels they are not getting their mail delivered on a timely basis. As the mailroom Supervisor, you ask one of your clerks to see if there isn't a faster method for sorting the mail. The clerk takes the assignment and reports two faster methods for sorting incoming mail. You adopt one method and all other clerks are instructed to begin using this faster method. Upper management notices the increased timeliness of the mail distribution and thanks you for your efforts.

As the Supervisor, you should follow up with the employee and let him/her know that (1) the new method has been adopted and (2) upper-management is appreciative of the increased efficiency resulting from the employee's contribution. Thus, the payoff in delegating comes in three ways: management is happy, you are happy, and the employee involved is happy.

TEAM BUILDING

The third technique is called team building. What is a team? Figure 5 – 9 gives us two of Webster's definitions (the first and the fourth) that bear some resemblance to what we mean by team building. Admittedly, Supervisors may sometimes feel that they are a draft animal harnessed to a vehicle, but this definition doesn't quite fit the typical workplace. Webster's fourth definition comes closer, but how would you distinguish between a number of persons associated together in work activity and a work unit, division, department, section or whatever? We clearly need a better description of a team if we want to determine what it really is.

The group of employees you supervise is your work unit. It can be made up of some very diverse people with often conflicting interests and needs. They are supposed to be team members, and you are supposed to be their team leader. Simple observation of the workplace shows that in many cases the work unit is anything but a team. There is often bickering, fighting, even undercutting of the common purpose. The common purpose is to get the work done in an efficient and effective manner.

In fact, your work unit is not your work team until, working together, you have collaborated to make it into a cooperative whole. Cooperation is one of three key elements of teamwork. The other two are respect and common purpose. Teams, therefore, must be built on these three key elements: cooperation, respect and a common purpose (Figure 5 – 10).

Here are some simple principles that you, the Supervisor, can follow to build an effective team:

- Treat all members of the team equally to avoid unhealthy competition or jealousy among staff. Avoid showing favoritism. Competition between team members is usually counterproductive. Competition between teams, on the other hand, is a frequently used motivation device in some companies. Cooperation, again, is the key element within a team.
- Talk with the team, at least occasionally, as a "team." This can be accomplished through staff meetings or informal conversations, where everyone is brought together. This collective approach establishes a basis for a common purpose. Remember, though, not to call a meeting unless you have something meaningful to discuss.
- Present a problem for the group to solve instead of always

Figure 5 – 9. Team—a Definition

Webster's Ninth New Collegiate Dictionary:
1. Two or more draft animals harnessed to the same vehicle or implement

OR

4. A number of persons associated together in work activity.

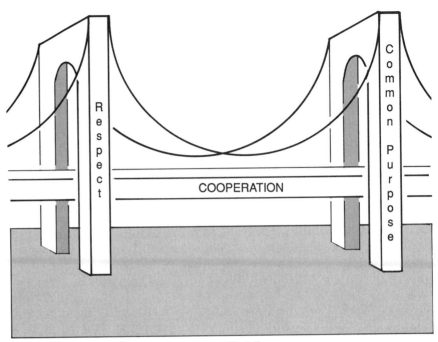

Figure 5 – 10. Key Elements of Team Building

solving it yourself or delegating it to only one or two people. This builds the sense of common purpose.

- Be sure everyone knows how he/she fits into the immediate department and how your department fits into the overall objective of the organization. This should be part of an initial orientation process, but it should also be informal and ongoing. You can do it through establishing a two-way communication system.
- Keep a steady flow of downward communication so your employees are well informed of what's happening in the organization.
- Allow for open, upward communication so employees know you are open and concerned about what they have to offer the organization.
- Always, always treat your team members with the same respect as you expect them to treat you (Figure 5 – 11).

HANDLING COMPLAINTS: GRIPES AND GRIEVANCES

Figure 5 – 11. Principles of Team Building

- Treat all members of the team equally
- Talk with the team
- Use group problem solving
- Be sure all members of the team know how they contribute to the whole
- Always treat your team members with respect

The fourth technique involves settling employee complaints. Supervisors have a very important role in determining the outcome of employee complaints, or, as they're often called, gripes and grievances. There are important differences between gripes and grievances, and both forms of complaints must be handled by the Supervisor. Before we get into the distinction between gripes and grievances, let's first look at them as one category: employee complaints. Complaints often can be easily resolved if handled appropriately and in a timely manner by the Supervisor. Effective communication, including that most important of all communication skills, listening, is again the key.

Active Listening

You will recall, we emphasized active listening in Chapter 4 on communication. Here, it becomes a critical part of the technique for handling employee complaints. Active listening is much more than just plain old everyday listening. Active listening means really tuning in to what an employee is saying. To effectively handle a complaint, you must use your active listening skills (Figure 5 – 12). How do you do this?

- Listen attentively and without distractions. If possible, go off to a quiet place where there will be no phone calls and no other employees.
- Avoid getting angry if the employee is angry or if you are personally being confronted or verbally attacked. Your anger will only compound the situation, and will probably block any understanding of what is causing the employee's anger. Isn't it hard to do? Of course it is. Chapter 11, Discipline in the Workplace, has some tips on how to accomplish the "stay cool" demeanor.
- Listen empathetically. Empathetic doesn't mean sympathetic. Being sympathetic means agreeing and taking sides; taking sides may be the completely wrong thing to do in the case of an unwarranted complaint. Being empathetic means you understand the other person's point of view, even if you don't agree with it. Empathetic responses: *Yes, Delores, I do understand how difficult this is for you.* Or, *Of course, Jasper, I really think it has been a very hard period for you, but. . . .* Or, *Marilyn, I didn't know you were getting a divorce. I'm sure that has made things tough on you.*
- Maintain eye contact with the employee. In this case, it demonstrates that you are taking the employee and the complaint seriously.
- Respond constructively and don't make light of the situation. If the employee thought it was funny or unimportant, why would he/she have brought it to you in the first place?
- Ask questions that help focus the conversation on the facts of the situation, instead of the emotional part. This is the most difficult element of active listening. In order to ask good questions, you must listen carefully. Always try to stick to the facts. *Yes, Tom, it is a difficult task. How do you propose to accomplish it?* Or *Tell me, Michael, exactly what did you see happen?* Or *Pete, you said everybody just got into a brawl. Such as who? Could you be specific?*
- Guide the conversation with questions, but don't interrogate and avoid making hasty remarks or offering instant solutions. Above all, be sincere.
- If necessary, postpone a solution in order to develop additional facts about the situation. Remember, there may be more to the situation than you and the employee know. Also, the complaint may involve more employees than just the one making the complaint.
- Finally, help the employee to resolve the situation or find a solution to the problem.

Figure 5 – 12. Active Listening Guide

- Listen attentively
- Avoid getting angry
- Listen empathetically
- Maintain eye contact
- Respond constructively
- Ask questions focused on the facts
- Guide, but don't interrogate
- If needed, postpone to develop additional facts
- Help the employee resolve the problem

Finding a Solution

Finding a solution to most complaints really involves repeating many of the steps involved in active listening. The biggest difference is that you must get the employee involved in wanting to find a solution, and then actually stating the solution to you. Sound impossible? Try these steps.

- Let the employee know that you are empathetic and want a solution just as the employee does.
- Review the facts with the employee (particularly once there has been an opportunity to "vent"). Paraphrase the facts as you understand them.
- Go over the situation until there is a mutual agreement on the facts as you have stated or restated them.
- Ask the employee questions that guide toward resolution, such as the following. *Do you have a solution in mind? What would you do if you were I? What can we do to keep this from happening again? What do you think is the best way out of this? How can we prevent this problem? What do you think would resolve the situation? What would make you feel better?*
- Have the employee state the solutions in his/her own words.
- As a last resort, offer a solution if the employee is really off track, but then have him/her restate the proposed solution in his/her own words.
- Using the employee's words, draw a conclusion. *So you agree that we should do. . . .*
- Finally, keep notes or documentation regarding the complaint and the agreed-upon solution. Even if the problem is seemingly resolved, it can come up at a later date. Your records, and the fact that those you supervise are aware that you keep them, provides some insurance of actually settling some problems (Figure 5 – 13).

Gripes and Grievances

Now we come back to the issue of separating complaints into their two traditional types: gripes and grievances. We need to look at gripes and grievances in two different contexts: human resource management where there is no labor agreement and human resource management where there is a labor agreement. Documentation for both union and nonunion situations should be the same.

Gripes and grievances without a labor agreement. In this human resource management context, gripes are complaints that allege a wrong or a problem that is not covered by the employee manual or management policy. For example, gripes often involve a perceived violation of fairness. These can be issues like the following:

- Jerry's desk is bigger than mine, but we're on equal jobs. It isn't fair.

Figure 5 – 13. How to Find a Solution

- Let the employee know you are empathetic
- Review the facts with the employee
- Go over the facts until there is agreement
- Ask questions to guide toward resolution
- Have the employee put solutions in his/her own words
- As a last resort, offer a solution
- Using the employee's words, draw a conclusion
- Keep notes

- How come you're always giving me the extra work assignments? Why not spread it around?

Grievances, on the other hand, allege specific violations of the employee manual or company policy, and typically are such things as:

- I did not get the right credit for vacation time this year.
- The fellows in the workshop keep whistling at me and making lewd remarks. Just because I'm the only woman in the shop doesn't mean they can treat me like that.
- How did Frank get promoted to Jim's job without letting the rest of us have a crack at it?

Handling gripes and grievances in a nonunion environment involves knowing the company policy and procedures and following the active listening process to solve the problems. In addition, many organizations without union agreements have an internal grievance/complaint-handling process to try to assure employees of fair treatment. As a Supervisor, you must know and follow that process.

Gripes and grievances with a labor agreement. In this human resource management context, gripes are complaints that allege a wrong or a problem that is not covered by the employee manual, management policy or the labor agreement. They would probably be exactly the same type of gripes as shown above!

Grievances, on the other hand, allege specific violations of the labor agreement. Again, grievances under the labor agreement would probably be exactly the same type as shown above, because labor agreements, employee manuals and company human resource management policy typically attempt to regulate the same sort of issues, although often very differently.

Under a labor agreement, you as the Supervisor must know the company policy and procedures and follow the active listening process to solve the problems (exactly the same steps followed in a nonunion setting). Supervisors should learn one or more problem-solving methods that they can use effectively. However, you must also familiarize yourself with the labor agreement, follow the formal grievance procedures spelled out in the agreement and keep a record of the dates when incidents and conversations took place, who was present or involved, a brief summary of the conversations and the result or resolution. This last step is critical because grievances under a labor agreement can sometimes be sent to an arbitrator, and it is up to you to have kept a record of the facts.

Essentially, then, handling complaints in a union or non-union environment still involves the same basic procedures for the Supervisor. After all, good supervision is good supervision (Figure 5 – 14).

APPLYING THE RULES

The final technique of supervision that we will cover involves how you carry out that familiar idea of the Supervisor as Rule Keeper. Here, we're examining that function in an active supervisory mode: applying the rules as a major form of supervision.

Know the Rules

Be sure you know what the rules are and the purpose of the rules. Supervisors who try to dismiss the rules as a formality that doesn't apply to them or their work unit, a let's-keep-this-among-us approach, will eventually find themselves in deep trouble. For example, what kind of conflict have they put themselves in when a company official sees that their group is violating the rules?

So it comes back to knowing and applying the rules. Even if you don't agree with all of the company rules, it is up to you as a Supervisor to communicate these rules to your staff and see that all of the rules are followed. These rules, in fact, provide the major basis for your ability to maintain order and discipline in the workplace and establish the concept of fairness in the workplace. They are the ingredients of critical importance to productivity.

Explain the Rules

Obviously, in order for you and your employees to follow the rules, you will have to explain the rules to them. NEVER, NEVER, NEVER assume that your employees know the rules and their purposes. It is your job, as Supervisor, to tell them about the rules and why they exist (Figure 5 – 15). To do this, you must:

- Communicate the rules and their purposes verbally and in writing to all your employees. Be comprehensive about this and take opportunities for refreshing their memories.

Figure 5 – 14. Separating Gripes and Grievances

GRIPES are complaints that allege a wrong or a problem that is not covered by the employee manual, management policy or the labor agreement. Gripes usually involve questions of "fairness" rather than violations of specific rules or regulations.

GRIEVANCES are complaints that allege specific violations of the employee manual, management policy or the labor agreement.

Both forms of complaints must be handled by the Supervisor.

Handling complaints in both a union and nonunion context involves the same basic procedures for the Supervisor. After all, good supervision is good supervision.

- Show your support of the rules. Say, *These are the rules and they exist for very good reasons.* Never say *I don't like it either, but that's just the rule.*
- Be an example by following the rules yourself. If you're above the rules, everyone else will either resent you or start ignoring the rules in the same way you do. For example, if you come in late every morning, chances are your employees will start coming in late every morning. If you then try to enforce the no tardiness rule, you will have a very difficult time establishing any credibility or compliance.
- Verbally present the rules to allow for questions or feedback. A discussion of the rules is critical to understanding them and why they exist.
- Follow up in writing. The work rules should be distributed to each employee and/or posted on a bulletin board. People can't be expected to follow rules they can't see.
- Communicate the outcome if rules aren't followed. Explain the organization's disciplinary process, the steps involved, the fairness principle embodied in the rules, the necessity for discipline and how discipline for infractions will be administered.
- Explain exactly how the rules will be enforced. Never leave yourself open to the I-didn't-understand-it-that-way routine.
- Enforce the rules as you have explained them. They will provide the basis for much of your ability to control the workplace. That is their ultimate use, and that is why they are such an important part of your group of techniques of supervision. With clear understanding of the rules, there is likely to be greater employee self-enforcement.

Many companies will require that you have employees sign-off on a statement acknowledging their discussion of the rules and opportunity to ask questions (Figure 5 – 16).

CONCLUSION

In this chapter we have examined five key supervisory techniques and how they are used by good Supervisors in human resource management. These techniques all involve communication and some of the important roles the Supervisor must carry out. Most importantly, they show how the Supervisor must function when dealing with employees and employee problems. The five techniques are:

1. Creating employee participation
2. Delegating responsibility
3. Team building
4. Handling complaints: gripes and grievances
5. Applying the rules

Creating employee participation is hard work, but it provides a significant basis for improved productivity and group effective-

Figure 5 – 15. Why Rules?

Organization rules and procedures:
- Form the basis for order and discipline in the workplace
- Establish fairness
- Are critical parts of productivity

Figure 5 – 16. Keeping the Rules
- Communicate the rules and their purpose
- Show your support of the rules
- Be an example by following the rules yourself
- Verbally present the rules and allow for questions or feedback
- Follow up in writing
- Communicate the consequences if rules aren't followed
- Explain exactly how the rules will be enforced
- Enforce the rules as you have explained them

ness. Delegating responsibility is a way for the Supervisor to work smarter instead of harder by having others do some of the work. Team building is an essential element in making sure that your unit works as a unit and minimizes internal competition and conflict. Handling gripes and grievances in a clear and effective manner through active listening assures your employees of fairness and understanding. Applying the rules in a clear and consistent fashion helps you to control the workplace and establish and maintain fairness in the treatment of those you supervise.

In each of these techniques, a Supervisor has to make a choice between talking to the troops or talking with the team. That choice becomes obvious: talk with the team.

Chapter 6

Supervising a Diverse Workforce

OVERVIEW

Today's workforce is made up of a group of people very different from the typical workforce of the past. Tomorrow's workforce will continue to diversify at a more rapid pace. The Supervisor, as the person Out in Front and in close touch with every aspect of the workplace, must confront the realities of a changing workforce. In this chapter, we will deal with what is involved in supervising this changing workforce, focusing on the following areas:

- How the workforce is actually changing, and what this means in terms of women, minorities, handicapped and the mix of younger and older workers
- The growth of a multicultural workforce, including different American cultures and a wider range of non-European immigrants
- Recognizing the importance of cultural differences and dealing with the foreign manager
- Supervising the multicultural workforce and finding some new ways to supervise in our diverse workplace

THE CHANGING NATURE OF THE WORKFORCE

We in the United States have always taken great pride in the idea that we have been a melting pot of people from all over the world. Of course, we also liked to pretend that all over the world only meant England and Europe, and some of us even got pretty fussy about which parts of Europe. But the reality is that our nation has always been made up of very many different kinds of people.

The first inhabitants, of course, were the many different cultures of Native Americans. Anthropologists tell us that even the Indians came here from Asia. Then maybe the Vikings, followed by the Italians (Columbus) and Spanish, French, English and Africans. Later immigrants included Europeans, Chinese, Latinos and, most recently, Southeast Asians. As the country kept expanding geographically, we also included more peoples like Hawaiians, Eskimos and Puerto Ricans.

We have always been a people made up of peoples. Today, however, the nature of our diversity is changing. Table 6 – 1 shows some of those changes. The number of blacks, Asians and Hispanics in our population has grown more rapidly than the number of whites. Look at the numbers and the percentages.

Table 6 – 1. The Changing U.S. Population

Population Shown in Millions

| | Actual | | | | Projected | |
	1972	**%**	**1986**	**%**	**2000**	**%**
Total Population	209.9	100.0	241.6	100.0	268.3	100.0
White	183.3	87.3	204.7	84.7	221.5	82.4
Black	23.6	11.2	29.4	12.2	35.1	13.1
Asian and other	2.9	1.3	7.5	3.1	11.6	4.3
Hispanic	*		18.5	7.7	30.3	11.3

* Not listed as a separate category
Source: U.S. Bureau of Labor Statistics

The image of a white, male, European workforce is disappearing (many would say it only existed in certain parts of the country for a very short period, anyway). The statistics in Table 6 – 1 show that the labor pool, those able and available for work, is actually shrinking because the population is also aging. The World War II "Baby Boom" adults are passing through middle age, and there are fewer people to replace them in the workforce. Since many minority populations have a higher birth rate than white populations, this means that the mix of peoples in the able and available workforce is changing even more rapidly than in the general population.

As a result of all these changes, the traditional white, male-dominated workforce is changing to include women, minorities, immigrants and the disabled. Since companies need people to fill in the spots being vacated by the Baby Boomers and the older workers, this trend will be continuing for a long time. It is a workplace reality. In the rest of this chapter, we will be dealing with how Supervisors can respond to this changing workforce.

Women in the Workforce

The U.S. Bureau of Labor Statistics predicts that by the year 2000, women will make up 47 percent of the workforce. Not all places will have as many women workers as that, but every place will have some. If you're a Supervisor, male or female, there are some special things you need to think about when it comes to supervising women in the workforce.

Supervising Women

Listed below are a few guidelines that will help you think about how to supervise women in the workplace (Figure 6 – 1).

Rule 1. Treat a woman as a person, an individual. Avoid thinking in stereotypes. What's a stereotype? One of those *Everybody knows that women. . . .* Have you ever noticed that nobody ever offers any proof or gets specific? Remember that behavior called bitchy or aggressive in women would be termed ambitious or masculine for a man. When you treat a woman as a person, she'll react with more confidence and improved performance when the basic ability is there—just as a man will.

Rule 2. Assume that women work as effectively as men. They can, they have, they do, intellectually and physically. Yet, women in the workplace are regularly challenged to prove themselves, particularly in physical skills, while men are simply assumed to have such abilities. Remember that we noted in Chapter 5 that internal team competition is usually unhealthy? Then why make some team members always have to prove themselves? The message it gives is that you have no confidence in them, and that leads to poor self-image and resentment. Is that how you want to treat your team? During World War II, American women performed virtually all jobs in our factories, foundries and farms. So, it's about time we drop the question of physical ability. Try walking up three flights of stairs with a bag of groceries in each arm and a 35-pound infant in a backpack. Try racing against an Olympic runner like Florence Griffith-Joiner, Jr.

Rule 3. Assume that women are at work to develop a career. Recognize the facts of life: women are the childbearers and may take time off for this, but having children does not make them any less effective or ambitious than men. In today's workplace, many women are choosing not to have babies. Many women start working after they've had babies, raised their children, are widowed or divorced. Many women return to the workplace after having babies, and an increasing number of fathers are taking care of babies or equally sharing in the process. These trends are not about to turn around. If anything, the percentage of women in the workforce will continue to increase.

This increase will have two main causes:

1. Companies will need more women to fill in the shortage of people in the labor pool. Therefore, providing some form of day-care assistance will increasingly become part of the corporate function with the growth of organization-sponsored day-care centers. This will happen out of necessity rather than good will or some social activism.

2. Just as men do, women will continue to need to work for a living to support themselves and their families.

Rule 4. Recognize women's strengths. Don't keep thinking of women as delicate, fragile things who can't make decisions. There are countless women who show great strength in coping with childbirth, childbearing, physical work, illness, financial adver-

Figure 6 – 1. When It Comes to Supervising Women...

THOU SHALT:

Rule 1. Avoid thinking in stereotypes.

Rule 2. Assume that women work as effectively as men.

Rule 3. Assume that women are at work to develop a career.

Rule 4. Recognize women's strengths.

Rule 5. Treat women with respect.

sity and death. Somehow, the fragile magnolia image of nineteenth-century Southern women is treated by some men as if it were reality in the entire country today. It never really was. Women worked alongside men in pioneering this country, often pulling the plow by back labor. Scarlet O'Hara may have looked fragile, but when necessity called, she was a tower of strength and determination. When it comes to strength, remember who outlives whom.

Rule 5. Treat women with respect. You must avoid sexual remarks and innuendos and enforce that policy with your employees. Using terms like Honey, Cutey, Babe, Dear, Doll and so on is demeaning, demoralizing and inappropriate in the workplace. How would a man feel if he were addressed that way by his Supervisor? Why would you assume a woman would feel any differently? Language that is acceptable between friends, lovers and spouses is not acceptable in the workplace. Because you didn't mean it that way doesn't mean others understand or appreciate your language. Above all, HANDS OFF! No touching, no sexual advances, no little pats. All of this stuff is sexual harassment, and sexual harassment is against the law.

Women as Supervisors

It is often difficult to go from peer one day to Supervisor the next. You have most likely had that experience yourself. One day, you were a worker, working alongside friends and colleagues. The next day, you were their boss. It is never easy to make that transition. It may be even more difficult to do this if you are a woman supervising a predominantly male staff. Why? Because many women (but not all) are culturally conditioned to take a secondary role to men. However, most women can get over this pretty quickly as they discover how really smart and capable they are. It may also be more difficult because many men and women still routinely violate all those guidelines we just listed and somehow act threatened when faced with a female boss. Here are a few tips for women working as Supervisors (Figure 6 – 2).

- Adopt and practice sound supervisory skills, and apply them universally, not according to gender. Forget about men and women and deal with team members. Good supervision is good supervision, regardless of gender.
- If you find your authority is being questioned or ignored, do the following. Find out if other Supervisors are having similar difficulties. It may be the environment or some other factor, not just that you are female. Confront the individual(s) with your observations just as we've told you to do with any other employee behavior problem. Outline expectations and the consequences if the expectations aren't met.
- If employees are going around you to your manager, take the following steps. Go to your boss and seek support to ensure that your authority isn't being undermined. If the problem persists, take the appropriate disciplinary action under organization policy.
- Note that these steps recommend treating problems re-

Figure 6 – 2. Tips for Female Supervisors

- Practice sound supervisory skills.

- Apply them generically, not according to gender.

- If your authority is being questioned or ignored:
  Determine if other supervisors are having the same problem.
  Confront the employee with your observations.
  Outline your expectations for a change in behavior.
  Explain the consequences if your expectations aren't met.

- If employees are going around you to your manager:
  Go to your manager and seek support to ensure that your authority isn't being undermined.
  If the problem still persists, take appropriate disciplinary action with the employee.

- Note that these steps treat the problem as you would treat any disciplinary problem in the workplace.

lated to gender differences exactly as you would treat any other disciplinary problem in the workplace.

Minorities in the Workforce

As shown in Table 6 – 1, minorities are becoming a larger part of the workforce. The U.S. Department of Labor's *Workforce Two-Thousand* report indicates a significant increase of minorities will continue to occur in the workforce. It is projected that the number of African-Americans will increase from 12% to 13%; Hispanics from 8% to 11%; and Asians from 3% to 4% by the year 2000. Because of affirmative action, government policies have encouraged hiring and promoting minorities. Because of a shortage of workers in the labor pool, organizations will, of necessity, continue that trend regardless of government policies. That is an economic fact of life in our capitalist society.

Supervising Minorities

For nonminority Supervisors, supervising minorities, like supervising women, may require some cleaning up of your act. Part of the process of change is in the realm of beliefs and part of it is in the way you behave. Realistically, neither we nor your employer can do much about how you think about minorities. Your beliefs are yours whether they are right or wrong. But how you behave in your job is under the control of your employer. Therefore, we're going to concentrate, as we do throughout this book, on appropriate behavior rather than appropriate thinking. We can only tell you how to be effective as a Supervisor in a diverse workforce.

Tips

Here are some behavioral tips, on your responsibilities as a Supervisor when dealing with minority-group workers (Figure 6 – 3).

- Be aware of any biases or stereotypes that you may have. While awareness is often the first step toward change, we're primarily concerned with raising your consciousness about any negative attitudes or biases.

- If you have any biases or stereotypes, do not act on them. This is essential to your role as a Supervisor. Remember the idea of fairness discussed in Chapter 5? Acting on any biases about minorities is a gross violation of that democratic workplace concept.

- Be sure you don't use any racial slurs or jokes, and ensure that co-workers don't either. If you're Irish, you may enjoy Irish jokes. But you won't like them if they're told to you by an Englishman. By using racial slurs you can also provide the basis for a very expensive antidiscrimination complaint. You know how much companies like Supervisors who do that.

- Don't treat minorities differently, either positively or negatively, just because they are minorities. Overcompensating by being too friendly to minority employees (just because they are minorities) will be seen as insincere and will breed resentment. Nobody enjoys being singled out for a characteristic that has nothing to do with who they really are.

- Avoid remarks like *Some of my best friends are...* If they really are, you won't need to mention it. If they aren't, your comments will only show your insincerity and ignorance.

- Don't show off your minority employees to try to prove you're not prejudiced. People want to be treated as employees. They like being seen as good employees, but they do not like being seen as token happy workers; nobody does.

- Provide for open communication so that minority employees are comfortable with you and vice-versa. Give positive feedback, and take corrective measures when necessary.

- If you're white, understand that the experience of being born Asian, black or Hispanic in America has in many ways been very different from your experience, while in other

Figure 6 – 3. Tips for Supervising Minorities

- Be aware of any biases or stereotypes you have.

- In spite of your biases, do not act on them.

- Don't use racial slurs or jokes.

- Don't treat minorities differently just because they're minorities.

- Avoid remarks like *Some of my best friends are...*

- Don't show off your minority employees to try and prove you're not prejudiced.

- Provide for open, comfortable communication.

- Understand that the experience of being born Asian, black or Hispanic in America is different.

- Don't overreact to charges of mistreatment or discrimination.

- Everyone has the same basic need to be treated fairly.

- Good Supervision applies to everyone.

ways just the same. These differences can be strengths in the workplace, if you're flexible enough to let them be.

- Try not to overreact to charges of mistreatment or discrimination. Becoming angry or emotional will only add fuel to the fire. Remaining calm can help to defuse the situation while you seek out a solution.
- Regardless of race or national origin, everyone has the same basic needs, one of which is to be treated equitably and fairly. And equitable and fair treatment does not mean always having to prove oneself.
- Finally, the principles of good supervision, as we are setting them forth in *Out in Front*, are the same for everyone who works in the modern workplace. Apply them.

Older Workers in the Workforce

The mix and range of ages in the workforce is changing very rapidly. Age 55 and over is the usual statistical breaking point used to define older workers. The U.S. Census Bureau estimates that by the year 2000, people 55 and over will have increased to 20% of the U.S. population. Youth will decrease from 20% to 16%. That means we are faced with an aging population and this is one of the reasons why we have been talking about the change in the workforce. There will be a labor shortage. The labor shortage will change a lot of things. We have already seen that because of pressure from employers to fill jobs, more women and minorities will be found in all aspects of the workforce. There will also be a graying of the American workforce.

The labor shortage will reverse the trend of past years on issues like early retirement. People will be encouraged to work to an older age. Some people may retire from one company and start another career. People will return to work from retirement. In fact, it's already happening. Walk into your favorite fast food restaurant and take a look at who's working there. It used to be high-school students. Now, you'll notice quite a few senior citizens. In states like Florida, with large senior citizen populations, many of the employees in the fast food and general service industries are seniors.

Supervising Older Workers

If you're a Supervisor in the 1990s, you'll be supervising a lot of older workers, regardless of your age. If you're younger than many of those you supervise (and that's increasingly likely), you may feel there are special problems involved, for you and them. If you are an older worker and a Supervisor, you may feel there is some difficulty in communicating across the age gaps, or in getting the right level of production out of your fellow employees. There are significant differences between older workers and others in the workplace. Of course, as with everyone else, there are many individuals who don't exhibit these differences. For the majority of people over age 55, however, there are some fairly common characteristics that will help us all to understand why they work and how they work (Figure 6 – 4). This understanding will help in determining how you can best supervise them.

Figure 6 – 4. Key Issues when Supervising Older Workers

- Recognize value differences
- Understand the impact of workplace changes
- Focus on job satisfaction
- Avoid stereotypes
- Appreciate experience

- Recognize value differences. Older workers have values that are often quite different from those of younger or middle-aged workers. The older workers will be products of the post-depression and post-World War II eras. They will probably have a strong work ethic, and will place a premium on job and financial security. In spite of the fact that they may have survived three or four mergers, acquisitions or reorganizations, they will probably be very loyal to the organization they work for.
- Understand the impact of workplace changes. They may not be used to the diverse workforce that exists today, having grown up when the workforce was predominantly white and male. They're going to have to get used to female, younger and minority Supervisors and workers, and different attitudes and values. You'll need to help them along through this process.
- While certainly not true of all seniors, many at this stage in their careers may put more value on how satisfying their current work is rather than salary increases or advancement. They may not desire or be motivated to advance any further, concentrating instead on job satisfaction.
- The older worker is probably not a "me first" type. Younger workers have grown up in the Me Generation. They may look for self-actualizing work and faster advancement and that may mean skipping from one organization to the next. The younger Supervisor may be a working mother. Not too long ago, this was an exception. Soon, it may be the rule. The older worker may have difficulty accepting these differences in values.
- Avoid stereotypes. If you dig into it, you'll find that the aging process has little to do with the core personality. The cranky old woman was probably once a cranky middle-aged woman. The tightwad old man probably watched his pennies throughout life. Again, the principles of good supervision will serve you well.
- Appreciate older workers as the valuable resource that they are, not as a liability because they are different from other workers. Studies have repeatedly shown that older workers generally possess some characteristics very much prized in the workplace: loyalty, lower rates of turnover, lower rates of absenteeism among healthy workers and higher rates of job satisfaction. One major study even found that older workers surpassed younger workers in both speed and skill, indicating that experience rather than age determined performance. Overall, most of the studies show that there are no great measurable differences in productivity between age groups in the workplace, although older workers generally have fewer on-the-job accidents, especially in situations that require judgment based on experience.

Some Facts about the Aging Process

The aging process doesn't necessarily mean the older worker will lose, or be losing, physical and mental abilities that will negatively impact work performance. Some may, but many won't. Don't assume there will be a memory loss or lost intelligence. There can be, but there generally isn't. What usually happens is that older workers need more time to learn new skills or adapt to a new situation. They can and will learn and adapt, but you need to allow more time at all levels of work. Even older executives need more time to learn new technology procedures. That need for more learning time is the single major workplace difference related to aging (Figure 6 – 5).

The following positive approaches to supervising older workers will help you more effectively supervise them (Figure 6 – 6).

- Recognize the value differences and overcome the generation gap through fair treatment and clear communication.
- Older workers may need retraining or orientation to new technology. In training, you may need to allow more time for completion. When new procedures are adopted, be patient. There may be a longer learning curve, but they do learn.
- Older employees may have less strength and endurance, may be less agile and may have slower reflexes than younger workers. However, the many years of experience, insight and wisdom the older worker has to offer certainly compensates for the physical differences.
- Status is often assigned not only by title but by years. Although the older worker should certainly be respected, don't forget that you are the Supervisor, and you are the one in charge. Don't hesitate to apply realistic performance expectations and to expect compliance with the work rules.
- Open communication should be fostered with the older worker, just like with other employees. It may be a new experience for them, but it will win out. It almost always does.

The Handicapped and Disabled in the Workforce

More and more organizations are coming to realize that the disabled are often an overlooked, but a valuable human resource. Through a combination of state and federal government affirmative action programs on behalf of the handicapped, organizations are more frequently finding effective and meaningful ways to employ handicapped workers in useful activities. (Although many people prefer the term *disabled*, the laws use the term *handicapped*. We, therefore, use both terms here.) Again, that now familiar labor pool problem we have been exploring in this chapter comes into play. If workers are scarce, you begin to hire people who may not be able to do everything but can do many things. In the long run, that economic fact of life may do more for employment of handicapped workers than any government program.

Physically and mentally impaired people, identified here under the legal definition of handicapped, are a large potential pool of very effective workers. Employers are also realizing that not everyone who has a high-school diploma or was born and raised in the United States can necessarily read and/or write

Figure 6 – 5. Myths versus Facts about Aging and Work

MYTHS:

Older workers will:

- Lose their physical abilities
- Lose their mental abilities
- Suffer major memory loss
- Generally decline in intelligence
- Be unable to adapt to new technology

FACT:

The single major workplace difference related to aging is that, usually, older workers may need a little more time to learn new skills or adapt to new situations.

Figure 6 – 6. Positive Approaches to Supervising Older Workers

- Recognize the value differences and overcome the generation gap.
- Provide time for retraining or orientation to new technology.
- Experience, insight and wisdom is just compensation for any less strength or endurance.
- Respect the older worker, but remember that you are the Supervisor in charge.
- Maintain open communication.

English. Many people who come out of our schools simply can't do that. In the Chicago area, for example, approximately 650,000 adults can read at only the fifth-grade level or below. Technically, they are handicapped.

Employing Handicapped Workers

If we are going to use this vast resource of people to advantage (both ours and theirs), we should think of some new ways to employ them. Here are some tips (Figure 6 – 7).

- Make things accessible. The best intentions in the world are thwarted if a person can't even get into the building! Many buildings are either inaccessible or largely unusable to many physically impaired workers. For example, a few simple ramps at the building entrance can suddenly open up a building to impaired workers at a relatively low cost. In new construction and rehabilitation of older facilities, planning and installation of ramps, wider elevators, sliding doors, wider toilet stalls and so on create accessibility.
- If reading or writing is not an absolute necessity to perform a job, don't make it a job requirement. Most companies have quite a few jobs where these skills are actually not necessary. Look for those jobs. If you are having a hard time filling a job, split off the nonreading/writing functions of several jobs and create a new job. For example, assume you have three jobs involving unloading and checking shipments. Unloading does not require reading; checking does. It may be possible to combine all the checking work into one position, and the person having that job would then assist the unloaders as time is available. The unloaders would not need to be literate.
- If possible, work with your organization's Employee Assistance Program (EAP) or Training Department to offer additional training or assistance to the functionally illiterate or to those for whom English is a second language.
- Physically disabled people have proven themselves to be contributing members of the workforce. There are many positions that they fill with no difficulty, and many more they can do with some difficulty but are more than willing to undertake. While they may need some special accommodations, for the most part they should be treated like other employees.
- Deaf or blind workers often have access to technical

devices that make it possible for them to do a very wide range of work activities.

- When dealing with handicapped workers, be patient but also be direct and specific when necessary. Set realistic limits but expect compliance.
- Be especially thorough when orienting and training the disabled worker so they can succeed.
- As much as possible, offer your assistance and support.
- Remember that they may be disabled or handicapped in some way, but not in their willingness to work.
- Aside from the particular constraints that might be imposed by the handicap involved, don't treat the handicapped worker any differently—positively or negatively—from any other worker.

THE MULTICULTURAL WORKFORCE

What is culture? It's all those things that make up a way of life and ways of doing things: language, customs, beliefs, feelings about work and play, methods of working and so on (Figure 6 – 8). Why should you be concerned with different cultures in the workplace? After all, this is America and people should act like Americans, right? Yes, no and maybe. Yes, we live in the United States and have always sought to develop a common set of values and attitudes, an American culture. No, we have never demanded full conformity, and have proudly retained many of our regional, ethnic and religious differences. Maybe we are becoming more alike through mass media and marketing, but there are still plenty of differences.

Some Differences in the Workforce

According to the U.S. Department of Labor's *Workforce Two-Thousand* report, immigrants are expected to represent the

Figure 6 – 7. Handicapped and Disabled in the Workforce

- Make things accessible.
- If reading or writing is not an absolute necessity don't require it.
- Work with your EAP for additional training or assistance.
- The physically disabled can fill many positions without difficulty.
- Deaf or blind handicapped workers often have access to technical devices that assist in work activities.
- Be patient, but also be direct and specific.
- Set realistic limits but expect compliance.
- Be thorough in orientation and training.
- Offer your assistance and support.
- Remember their willingness to work.
- Treat handicapped workers the same as everyone else.

Figure 6 – 8. What Is Culture?

Culture is all of those things that make up a way of life and ways of doing things: language, customs, beliefs, feelings about work and play and methods of working.

largest share of the increase in the workforce and population as a whole in the next few decades. In spite of new and tightening restrictions, approximately 600,000 legal and illegal immigrants will enter the United States each year between now and the year 2000. For the decade of the 1990s, that amounts to an additional 6 million people, more than the population of Los Angeles or Chicago. In the past 15 years and for the foreseeable future, most of these immigrants have and will come from countries and cultures other than European. Some of these cultures are Afgani, Bahamian, Cambodian, Chileno, Chinese, Cuban, El Salvadoran, Eritrian, Ethiopian, Filipino, Greek, Haitian, Indian, Iranian, Iraqi, Jamaican, Japanese, Korean, Mexican, Pakistani, Thai, Turkish, Vietnamese, etc. (Figure 6 – 9).

Understanding Cultural Differences in the Workplace

Today's workplace is made up of people from our previous groups of immigrants, such as African, Belgian, Canadian, Dutch, English, French, German, Hungarian, Irish, Italian, Mexican, Polish, Spanish, etc., plus those mentioned previously and many others. Of course, this varies in different parts of the country, just as it always has. In factories, it is common to find a very wide variety of ethnic backgrounds, with many first generation immigrants. In high tech labs, one often finds immigrant Asians and Indians who have specialized in computer technology. In previous eras, we find that whole industries were dominated by certain immigrant groups: Irish and Chinese building the railroads, Slavs in the steel mills, or, going way back, French in the fur trade. All of these people, then and now, brought their cultures with them, practiced and often maintained those cultures while melting into the American culture.

These are the people whom you will be supervising. If you don't understand some aspects of their cultures, you will have a difficult time doing your job.

Examples of Cultural Differences that Affect Work

When we talk about cultural characteristics, we talk in generalities (Figure 6 – 10). There will always be exceptions to the generality. Those we discuss here are generally recognized as common cultural traits, even though individual members of a given culture may not follow all of the practices. Thus, you can only use these descriptions as a guide to understanding behavior. Do not assume that they are true of all members of any group. Avoid turning useful concepts into stereotypes.

Some foreign-born employees may be reluctant to demonstrate initiative. Their respect for authority and fear of

Figure 6 – 9. Immigration: The Future Is Clear

Immigrants will be coming from cultures such as Afgani, Bahamian, Cambodian, Chileno, Chinese, Cuban, El Salvadoran, Eritrian, Ethiopian, Filipino, Greek, Haitian, Indian, Iranian, Iraqi, Jamaican, Japanese, Korean, Mexican, Pakistani, Thai, Turkish, Vietnamese, etc.

Figure 6 – 10. Examples of Cultural Differences

- Some foreign-born employees may be reluctant to demonstrate initiative.
- Many Asian cultures are sensitive to group harmony.
- Cambodians are teamworkers.
- In the Middle East, one's word is one's honor.
- To Mexicans, small talk before business is the way to do business.
- French workers may have difficulty believing working hard will improve one's position beyond what one is born into.
- To Japanese, group success rather than individual success is what counts.
- Japanese and Europeans feel that employees will be motivated simply by being hired by the company.
- Don't wear a white dress or suit to a Chinese wedding. (Of course, Miss Manners says guests shouldn't wear a white dress at an American wedding either!)

failure can make them hesitant to do so. They may pretend to understand direction because they are afraid to ask questions. East Indians, for example, expect decisions to be made by their superiors in the company and have a difficult time with participative supervision and management.

Many people from Asian cultures are sensitive to group harmony. As a result, being singled out for praise may not be viewed as positive. Praise can actually be interpreted to mean something is actually wrong (the exact opposite of what you may have intended). Tangible rewards may be viewed as more satisfactory. Japanese, for example, focus on group or team effort, with a great respect for seniority when it comes to promotion. Singling out an individual for praise is unacceptable. Promoting a junior employee over others, regardless of ability, is unacceptable. To Japanese, tasks that involve working alone are not going to be very satisfying.

Cambodians are teamworkers, and their usual team is an extended family unit (fathers, mothers, sons, daughters, aunts, uncles, cousins). Several food processing plants have had the experience of hiring these teams (usually without understanding they were relatives). The Cambodian team has an apparent leader and that is the person with whom the Supervisor must deal. (Usually the best English speaker acts as if he is the leader—it would never be a woman.) The actual leader, however, may be an older and less physically able team member who doesn't speak any English but commands great respect from the team. These teams are usually very productive, outperforming most other teams. At the same time, they often carry some very unproductive family members along with them, one of whom might even be the team leader. Supervisors who try to do something about the unproductive members will, quite simply, either lose the whole team or destroy its productivity and morale.

In the Middle East, one's word is one's honor. If an employee is instructed to improve a certain element of his work and then the conversation is summarized in a memo, the employee may take the putting of a commitment in writing as a symbol of mistrust. On the other hand, Americans often assume a commitment exists with a handshake or smile. Middle Easterners do not. They must specifically state *You have my word* or the clear equivalent.

In the Mexican culture, it is often common to engage in small talk before dealing with the business at hand. Informal conversation before giving directives is viewed as the proper way of working together. Supervisors who ignore small talk and come right to the point are seen as crude, rude and demeaning. Most Hispanic cultures reflect these same values, but there are great variations. Argentinians and Cubans, for example, seem to get to business much more quickly than, for example, Mexicans or Puerto Ricans, who place greater value on developing a relationship.

The French are very class conscious, and this carries over into the workforce. French workers may have difficulty accepting the American idea that working hard may improve one's position beyond the place in life one is born into. That place in life is usually seen as reaching back many generations.

Asians generally have a great deal of concern with face. Westerners often refer to this as saving or losing face, but the concept also includes gaining, maintaining and having face. For example, *Santosumi san has much face* means Mr. Santosumi is a person who has a sense of his oneness and place in life, but the phrase does not imply status or intelligence. If you treat Mr. Santosumi in a way that might embarrass him, for example, by singling him out for praise or chastisement, he will lose face, and so will you!

Both Japanese and Europeans feel that a key employee will be motivated by the mere fact he or she has been hired by the company. In Japan, particularly, this motivation through identification with the company may go on for a lifetime of very low wages. In the United States we may look more at monetary rewards as motivational or as a form of recognition.

If you're going to a Chinese employee's wedding, don't wear a white dress or suit. White is the color of mourning for the dead.

When it comes to safety, take special care. Many cultures do not place the same emphasis and value on individual life as we do. Many do not place legal responsibility for safe procedures on the employer. Consequently, in many parts of the world, unsafe practices tend to be the rule rather than the exception. When you have an employee from one of these areas (for example, much of Southeast Asia and much of the Middle East), you will have to watch his/her safety habits and probably do a good deal of retraining.

The above information is not intended to serve as a comprehensive list. It doesn't even touch the surface. It is intended to demonstrate the impact that cultural differences can have in the workplace.

Supervising for a Foreign National Manager

Working for a foreign national manager is going to become more and more common for the Supervisor. The growth of foreign-owned or mixed-ownership companies in the United States is increasing rapidly. Most Americans do not understand that the same is true all over the world. Americans see this change as Japanese or Arabs buying up American companies. Actually, the largest foreign purchasers of U.S. companies have been Canadians, British, Dutch and Japanese, in that order. The rest of the world sees American and Japanese companies, in that order, buying up everything! The reality is that the world's economy has truly become global. There are few finished products made anywhere in the world that do not involve a significant number of parts made in other countries.

Moving Managers

One of the natural results of this is that the home-country headquarters of a company frequently send trusted managers to newer operations. As these companies become truly global, they ship their managers anywhere they're needed. Many truly global companies require their managers to take assignment outside the manager's home country if they wish to proceed to the higher levels in the company. Thus, it would be relatively common to find an American-based multinational company sending one of its British managers to work in North Africa, or even in the United States. So the odds continue to increase that someday you will be working for a foreign national manager.

Special Problems for the Supervisor

Managing a company in a country other than your own is a difficult business. It is very similar to supervising a multicultural workforce, only this time you, the Supervisor, are on the receiving end (Figure 6 – 11). The problem for you is twofold:

1. To remind yourself that the foreign manager may not really understand the American culture and, therefore, may make all sorts of mistakes.
2. To figure out what you can do about that so the manager's expectations can be met, or if they cannot, the reasons will be understood.

To illustrate the special task you have, let's look at an example.

CASE STUDY: FOREIGN NATIONAL MANAGERS

Seiji Sakamoto is from Japan. He has been a manager for 12 years in Japan and Southeast Asia for the Nihon Manufacture and Export Consortium (NMEC). This is his first assignment in the United States. NMEC bought out your company and has introduced all kinds of new human resource management techniques. They have introduced common employee uniforms for everybody, including the managers. There is a common cafeteria where everyone eats together, officers and employees. Teams and quality circles have been established, and top management has set a goal of increasing

Figure 6 – 11. The Supervisor and the Foreign National Manager

The Supervisor should assume that the Foreign Manager:

• Knows what he is doing but may not know how it will work with Americans

• Regards his request as reasonable so your job is to suggest better ways to achieve the goal

• Is trying to adapt to the American business style just as you are trying to adapt to his.

---

production by 20% in the next 12 months. They advise you that this will still leave the U.S. plant production 16% behind a comparable NMEC facility in Osaka.

Your team has not been achieving the production goal. Mr. Sakamoto advises you to introduce group criticism sessions as a way to identify problems and improve production. He insists on attending them with you at the beginning in order to get things off to a good start.

At the first session, Mr. Sakamoto asks your employees, including you, to stand up and describe what they are doing wrong, what mistakes they have made in the past week, and then let the group address how these mistakes and errors should be corrected. Nobody bites.

What's going on here? There is a major problem of cultural misfit. In the Japanese culture, self-criticism is a way of life. If you say to a Japanese woman, *Your husband is a very smart man, I am impressed by how well he does his job,* she will likely respond, *Oh no, you must be mistaken; my husband is not so bright and only does what is necessary.* This is the sort of thing that goes on throughout Japanese life. In the Japanese culture, praise can only be responded to with self-criticism, and the self-criticism is sincere. Thus, in a typical Japanese employee self-criticism session, workers will stand and recite their individual mistakes and ask for suggestions. The basic assumption is that you are prone to mistakes and have many shortcomings as an individual.

It would take a lot of conditioning before that could become routine in an American factory, and we doubt if it ever really could. This is not to say American workers can't find ways to improve things. They can, but they prefer to do things in a positive way. How can we improve the process? What else can be done to make things better? That is the way we do things in our culture. The basic assumption is that we are doing things well, but will try to improve.

So what do you do in these circumstances? You probably didn't know about the self-criticism part of the Japanese culture, but you would know that Mr. Sakamoto's approach was not going to be successful. Therefore, you try to head it off with some explanations about how you think things could be improved with your work team.

Essentially, you should assume that the foreign manager:

- Knows what he (99% of the time, since affirmative action is not alive and well outside the United States) is doing, but may not know how it will work with Americans.
- Regards his request as reasonable, so your job is to suggest better ways to achieve the goal.
- Is trying to adapt to the American business style just as you are trying to adapt to his.

In the next section, we deal with some tips on supervising the mixed-culture workforce. Many of these tips can be applied upward to your foreign manager.

Supervising the Mixed-Culture Workforce

So how are you going to put it all together? The cardinal rule is be aware of cultural differences and the fact that your style of supervision can be misinterpreted due to cultural differences. Don't assume that just because they're in America now, they understand all of the American ways (Figure 6 – 12). The following tips will help you supervise employees of other cultures:

- Be aware of any of your own biases and remember that it is your responsibility to treat all employees fairly. Fairness is a concept that can cross most cultures.
- Simplify your communication. Use clear and precise writing and instructions. Use universal symbols and drawings or photographs, when possible.
- Be specific and clarify if words or phrases are misunderstood, especially where safety rules and instructions are involved. Insist on feedback to assure that you are understood. If they cannot use English, always use an interpreter and get the feedback through the interpreter. Strongly encourage interaction so you can check to be sure that you are being understood. This will be difficult with some cultures, but try.
- Avoid using slang or jargon that could be misinterpreted or not understood. How would you explain *Lucked out* or *How come?* An Englishman once said about the American/British relationship: *We are separated only by a common language.*
- Avoid jokes or sarcasm. Some humor does translate from one language/culture to another. Much does not. Worse yet, some translates in exactly the opposite meaning.
- Look at the employee's nonverbal clues to check understanding. If he/she looks confused, inscrutable or blank, someone missed the mark.
- Do not talk down to the foreign-born employee.
- As possible, show some interest in their culture, language and traditions (Figure 6 – 13).

Figure 6 – 12. Putting It All Together: The Cardinal Rule

Be aware that cultural differences exist and that your style of Supervision may be misinterpreted due to cultural differences.

Handling Cultural Conflicts between Co-Workers

As the Supervisor, you must recognize that it is your responsibility to be aware of any conflict and try to resolve it. As we have already emphasized in Chapter 3, you can't just ignore conflict. You must confront it and resolve it. Cross-cultural conflicts between employees can flare up and become very hot, often resulting in racial or ethnic slurs and leaving bitter memories. There are some straightforward techniques you can use to correct these situations:

- Avoid taking sides, regardless of your own prejudices or preferences, and stress similarities rather than differences.
- Maintain an unemotional stance. Don't overreact by offering ultimatums. Transferring or firing an employee is usually not an effective solution. Remaining levelheaded helps to defuse otherwise emotionally charged situations.
- Give the employees an opportunity to resolve their differences, with you as a neutral, objective, third party. In this mediator role, use the active listening process to identify the facts and attempt to move everyone toward a solution.
- If necessary, refer the individuals for further coaching or training, usually to the Employee Assistance Program or Employee Relations in your organization. Remember, you are not a counselor.

CONCLUSION

We have examined some aspects of the American workplace of today and the immediate future. Some may not be comfortable with the vision, but all are going to have to live with it. In reality, our nation has gained from each immigrant group that has come to us. What is American food if it isn't roast beef, pizza, ribs, tacos, egg rolls, spaghetti, sauerkraut, sushi, and so on? What are Americans if we aren't the vast melting pot we have always been?

Figure 6 – 13. Tips on Supervising across Cultures

- Be aware of any of your biases.
- It is your responsibility to treat all employees fairly. Fairness is a concept that can cross most cultures.
- Simplify your communication.
- Be specific and clarify if words or phrases are misunderstood.
- Insist on feedback to assure that you are understood.
- Avoid using slang or jargon.
- Avoid jokes or sarcasm.
- Get feedback.
- Strongly encourage interaction.
- Look at the employee's nonverbal clues.
- Clarify any miscommunications or misunderstandings.
- Do not talk down to employees.
- Show some interest in their culture, language, traditions.

We have also gained greatly from the efforts of women and minorities, who have always worked but are only now coming into positions of authority and recognition in the workforce. And when you as a Supervisor are puzzled over a current problem, you will most likely turn to the senior worker for comments or advice. Slowly, but nonetheless steadily, we are also recognizing and sharing pride in the ability of handicapped workers to play a meaningful part in the workplace.

The workplace of today and tomorrow will be made up of more minorities, more women, older workers, handicapped workers and new citizens from other lands who, by their presence, affirm their belief that America is still the best place to live, work and raise children. Supervisors will need to be sensitive to all of these changes, and to the different ways one must deal with all these different people who make up the workforce.

At the same time, we must recognize the need to maintain control and to supervise. Good supervision and fairness are the keys to success with the diverse workforce.

Chapter 7

Supervisors and the Law: Employment and Workplace Law

OVERVIEW

In this chapter we deal with some of the more difficult issues facing Supervisors today:

- Complying with employment or workplace law
- Assuring your employees comply with the laws
- Sending a consistent message that the law is the law and must be respected

We will review key aspects of two kinds of laws with which you must be familiar: employment law and workplace law. *Employment law* is the term generally used for federal and state laws that regulate possibly discriminatory employer behavior toward workers. This covers such laws as:

- Title VII of the Civil Rights Act, 1964
- Pregnancy Discrimination Act, 1978 (amendment to Title VII)
- Civil Rights Act, 1964 (EEOC)
- Age Discrimination in Employment Act, 1967 (ADEA)
- Rehabilitation Act, 1983
- State laws dealing with discrimination in the workplace

Workplace law is the term generally applied to federal and state laws that regulate employer behavior in regard to democratic rights of workers; the relationship between the employer, employee and labor organization; and the relationship between the employer and employee(s). This covers such laws as:

- Williams-Steiger Occupational Safety and Health Act, 1970 (OSHAct)
- OSHA Hazard Communication Standard, 1988 (HazCom)
- Fair Labor Standards Act, 1938 (FLSA)
- Equal Pay Act, 1963 (an amendment to the FLSA)
- National Labor Relations Act (NLRA)

There are many more workplace laws that impact on the employer and the employee but that are not, generally, within the province of the Supervisor. For example, the minimum wage laws on federal and state levels are obviously important, but Supervisors do not usually set wages. These laws have also been in place for many years and are usually treated as a given. Therefore, with the exception of the more recent equal pay provisions, we have not dealt with them. Various pension laws and regulations are in this same category.

This chapter, then, will deal with only those laws that are relevant to your job responsibilities. We will describe why the laws exist, the impact of the laws and your role in complying with and enforcing them.

WHAT'S INVOLVED, AND WHY BE CONCERNED?

First of all, you must be familiar with key aspects of these laws because your employer expects that. It's a very important part of your job. If you violate the employment or workplace laws, or allow those you supervise to violate them, you are exposing your employer to potential financial liability and damage to reputation. You may also have some personal liability if you or those working under your supervision commit acts declared illegal by employment or workplace laws and a complaint or lawsuit is filed.

Employment Law Background

Numerous employment laws have been passed by federal, state and local governments over the past several decades that seek to stop and to correct various forms of discrimination against specific groups or classes of people. These groups are collectively referred to as *protected classes.* The laws generally prohibit employment practices that discriminate based on race, religion, national origin, gender, age or status as a minority, veteran or disabled person (Figure 7 – 1). These protected classes have been selected for special focused protection under the law because the courts and the governments have found that these particular people have been or are being discriminated against in employment.

Workplace Law Background

Workplace laws have generally been passed by governments because it is in the public interest to assure employees of the following:

- Minimum employment conditions regarding wages, hours, benefits and working conditions
- Safe working conditions regarding practices, materials, equipment, processes and procedures
- The right to representation by a labor organization of the employee's own choosing

Governments generally intervened to impose workplace laws because they found that employers were not uniformly providing these conditions, that the employer/employee or the employee/union relationship needed regulation.

Since governments in our country are elected by the people to represent the people, and since a majority of the legislative

Figure 7 – 1. It's the Law

You May Not Discriminate in Employment Practices Against Any Protected Classes

That means, NO DISCRIMINATION based on:

- Race/color
- Nationality
- Sex
- Religion
- Age (anyone 40 and older)
- Mental or physical handicap (nondisqualifying)
- Veteran status

bodies had to vote in favor of these laws, our employment and workplace laws do reflect the wishes of most of at least a majority of the people. Courts only step in to interpret these laws by applying the statutes to individual cases that come before them. Courts do not write those statutes. Employment and workplace law is present at the federal, state and local levels. Supervisors need to know what the laws dictate and how to abide by them. Simply put, the law is the law and it's the Supervisor's job to see that it is followed in the workplace.

Issues of Morality

The law is the law does not really go far enough if we are going to effectively live with our workplace and employment laws. Most employers and Americans would say: The law is the law, and it's right! The workplace and employment laws are designed to protect employees (and ultimately, everyone is an employee, including top management), not to make life difficult for Supervisors and employers. The laws are designed to prohibit past and even present wrongdoing by employers. This does not mean that all employers are guilty of such practices, although many were and some still are.

We do need to recognize that these laws arose out of practices such as:

- Endangering children's health and preventing their education by working them long hours in unsafe conditions at below standard rates of pay
- Setting up employer-controlled unions to prevent employees from organizing their own union
- Paying minorities and females less for doing the same job as white males
- Hiring discrimination practices based on race, creed, national origin and sex
- Terminating older workers just because they were old or because the employer could hire younger workers for a lower salary or wage
- Endangering employees through unsafe working conditions or failure to inform them of hazardous conditions or chemicals

Thus we arrived at our present condition.

EMPLOYMENT LAW

We will focus in this section on the five federal employment discrimination laws every Supervisor should know. We will also discuss some representative state laws that go beyond the federal laws.

Title VII of the Civil Rights Act, 1964

This is the paramount employment law. It defines what employers are prohibited from doing in the area of the civil rights of protected classes. It seeks to prevent any discriminatory practice based on race, color, religion, sex or national origin. Under

Figure 7 – 2. Title VII of the Civil Rights Act, 1964

An employer may not use a person's race, color, religion, sex or national origin as a basis for:

- Not hiring an applicant for employment
- Discharging or disciplining an employee
- Setting an employee's wages, benefits or working conditions
- Dealing with an employee or job applicant in a way that would deprive him/her of a job opportunity
- Discrimination against an employee or job applicant because he/she opposed an unlawful employment practice

Title VII, the employer commits an unlawful employment practice if the employer or an agent, such as a Supervisor, uses a person's race, color, religion, sex or national origin as a basis for:

- Failing or refusing to hire an applicant for employment
- Discharging or otherwise disciplining an employee
- Determining an employee's compensation, including fringe benefits or other terms, conditions or privileges of employment
- Limiting, segregating or classifying an employee or an applicant for employment in a way that would tend to deprive him/her of an employment opportunity or otherwise adversely affect his/her status as an employee

It is also unlawful to discriminate against an employee or applicant for employment because he/she has opposed an employment practice unlawful under Title VII, or filed a suit or complaint under Title VII.

Some employment practices can reflect an old-school gentlemanliness that is simply against the law. For example, giving traveling assignments only to a man because it might be difficult on the woman's family is discriminatory. (What about the man's family?) More importantly, however, travel experience is often a requirement for advancing to higher-level jobs. Thus, a sweet thought becomes not simply dated; it's discriminatory.

Other discriminatory employment practices can simply reflect a subconscious bias or prejudice of a Supervisor or employer. For example, don't have a rule that ethnic employees can't speak their own language (unless you can prove it actually affects operational efficiency or safety). The rule, on its face, singles out the group involved and deprives them as a class (Figure 7 – 2).

Pregnancy Discrimination Act, 1978

This law is an amendment to Title VII and was adopted because many employers denied medical benefits and leave for pregnancy. The law has three very straightforward key principles that must be applied in the workplace (Figure 7 – 3):

1. A pregnant woman should be treated the same as any other applicant or employee.
2. Pregnancy is to be treated the same as any other temporary disability.
3. A pregnant woman need not be given preferential treatment.

Civil Rights Act, 1964

Figure 7 – 3. Pregnancy Discrimination Act, 1978 (Amendment to Title VII)

Three Rules To Be Followed:

ONE: A pregnant woman should be treated the same as any other applicant or employee.

TWO: Pregnancy is to be treated the same as any other temporary disability.

THREE: Pregnancy need not be given preferential treatment.

Much more than Title VII was included in the Civil Rights Act of 1964. It is the basic law that seeks to correct a number of wrongs relating to voting rights, public access rights, education, etc. In addition to the Title VII provisions we have described, the Civil Rights Act also prohibits harassment in employment because of one's sex, race, color, religion or national origin. Currently, sexual harassment is a growing issue although all other forms of employment harassment of protected classes are covered under the same mandate, such as race, color, religion, national origin and age. Sexual harassment, while not specifically mentioned in the law, has been included in the mandate as a result of judicial interpretation. Putting it simply: it is unlawful to harass a person at work or in connection with work because of or with regard to sex, race, color, religion, national origin or age (Figure 7 – 4). See the section on sexual harassment later in this chapter for further discussion.

Age Discrimination in Employment Act, 1967

The Age Discrimination in Employment Act of 1967 (ADEA) essentially extends Title VII coverage to the protected class of age (those 40 and over). Under ADEA, employers or their agents, such as Supervisors, commit an unlawful employment practice if they use an individual's age as a basis for:

- Failing or refusing to hire an applicant for employment
- Discharging or otherwise disciplining an employee
- Determining an employee's compensation or other terms, conditions or privileges of employment
- Limiting, segregating or classifying employees or applicants for employment in a way that would tend to deprive them of an employment opportunity or otherwise adversely affect their status as an employee
- Reducing the wage rate of an employee in order to comply with other requirements under the ADEA
- Discriminating against employees or applicants for employment because they opposed an employment practice unlawful under the ADEA, or because they filed a charge, testified, assisted or participated in an investigation, proceeding or litigation under the ADEA
- Selecting younger employees by employment notices or advertisements indicating a preference limitation or specification based on age (Figure 7 – 5)

The Rehabilitation Act, 1983

The Rehabilitation Act of 1983 deals with many discriminatory practices against disabled people and establishes basic rights, including public access to facilities and the opportunity to do a job. In the workplace, the law calls for Affirmative Action for the disabled. It defines disabled as anyone who (1) has a physical or mental impairment, (2) has a record of such impairment or (3) is regarded as having such impairment. This is one of the newer areas of law, and there are still problems in matching existing abilities of disabled workers and available jobs. Most of those problems are with lack of knowledge on the part of the employer as to just how much disabled workers can actually accomplish. There are some very clear-cut rules for employers and their Supervisors to follow.

- DO evaluate disabled applicants in terms of accommodations, such as modifications in equipment, that can be made to enable them to fill the job.

Figure 7 – 4. Civil Rights Act, 1964

It is against the law for

an employer
an agent of an employer (such as a Supervisor)
a fellow employee

to harass any employee because of or with regard to:

Sex
Race
Color
Religion
National origin
Age

Figure 7 – 5. Age Discrimination in Employment Act, 1967 (ADEA)

YOU CANNOT:

- Fail or refuse to hire an applicant for employment because of age.
- Discharge or discipline an employee because of age.
- Base an employee's compensation or other terms, conditions or privileges of employment on age.
- Limit, segregate or classify employees or applicants for employment in a way that would tend to deprive them of a job or adversely affect their status as an employee because of age.
- Reduce the wage rate of an employee in order to comply with other requirement under the ADEA.
- Discriminate against employees or applicants because they opposed an employment practice unlawful under ADEA or filed a charge, testified, assisted or participated in an investigation under ADEA.
- Place an employment notice or advertisement indicating a preference limitation or specification based on age.

- DO see each disabled applicant as an individual with a unique combination of strengths and weaknesses and try to match this combination with a job in your unit.
- DO consider redesigning a job so a disabled worker can do it.
- DO give newly hired disabled employees close attention at first to see how they're adjusting to the job and then make job modifications, if necessary.
- DO let disabled employees know you expect a full day's work, just as you do from any other employee.
- DO consider disabled employees for promotion, just as you do other employees.
- DO take steps to promote acceptance of disabled employees by the other workers in your unit. For example, urge them to include disabled co-workers in coffee breaks and other social activities.
- DO consider a disabled applicant in terms of actual job requirements.
- DON'T refuse to hire a disabled applicant who is otherwise qualified for the job.
- DON'T automatically disqualify an applicant with a disability.
- DON'T use rigid physical requirements that would disqualify disabled applicants, unless they're actually job-related.

State and Local Laws

All states, most major cities and many large counties have adopted stringent employment laws generally patterned after the federal laws, but often extending rights to areas not covered by the federal laws. Sometimes, these laws even create additional rights or benefits for protected classes or add additional penalties against employers for violations of the laws. We cannot possibly cover the provisions of the laws in 50 states and hundreds of cities and counties. However, we have chosen two examples to show the range of state activity in employment law: Illinois and California.

Illinois Employment Law

Illinois has a Human Rights Act that spells out many protections for employees. Essentially, the act is modeled on the amended U.S. Civil Rights Act, although it adds some specific areas intended but not specified to be covered under that Act. An employer commits an unlawful employment practice under the Illinois Human Rights Act if the employer considers a whole series of things in connection with employment decisions. Specifically, employers may not (1) refuse to hire on the basis of or (2) segregate or act without respect to any protected classes in any of the following employment actions:

- Recruitment
- Hiring
- Promotion
- Renewal of employment
- Selection for training or apprenticeship
- Discharge
- Discipline
- Tenure
- Terms, conditions or privileges of employment

Illinois law prohibits discrimination or employment decisions based on:

- Race
- Color
- Religion
- National origin
- Ancestry
- Age
- Sex
- Marital status
- Disability
- Unfavorable discharge from military service
- Sexual harassment

If you compare these terms to the laws illustrated in Figures 7 – 1 to 7 – 5, it is apparent that there are a number of additional categories, such as marital status and unfavorable discharge. That extension of coverage and detailing of categories is fairly typical of state employment laws. Illinois, of course, also has many other laws dealing with employment issues.

California Employment Law

California has a wide range of employment laws, most of them patterned after the federal laws, but significantly extending or specifying employee rights. To illustrate this, we will focus on the California law regarding sexual harassment. California prohibits

Figure 7 – 6. What Is Discrimination?

It's legal to discriminate *among* employees; it's *not* legal to discriminate *against* employees.

In hiring, promotion and other areas of employment, you may discriminate between employees on the basis of:

SKILLS
EDUCATION
EXPERIENCE
ABILITY
PERSONAL PREFERENCE

But this discrimination cannot be based on:

RACE
COLOR
NATIONAL ORIGIN
SEX
RELIGION
AGE
MENTAL OR PHYSICAL HANDICAP
VETERAN STATUS

Figure 7 – 7. Discrimination in Employment

Discrimination: actions against protected classes that treat them disadvantageously and differently from other people.

sexual harassment in employment and defines harassment with the following extensive detail. Harassment is:

1. Verbal harassment, such as epithets, derogatory comments or slurs
2. Physical harassment, such as assault, impeding or blocking movement, or any physical interference with normal work or movement
3. Visual forms of harassment, such as derogatory posters, cartoons or drawings
4. Demand for sexual favors, such as unwanted sexual advances that condition an employment benefit upon an exchange of favors

That's pretty clear, isn't it? It is important for a Supervisor to be aware of state and local laws and regulations with regard to employment. The combination of federal, state and local laws that regulate the protection of classes of employees means they are really protected, and it is part of your job to see to it that the laws are respected and obeyed by yourself and those you supervise.

What Does All of This Mean?

Does it mean discrimination is illegal? Not exactly. You and your employer have the right to discriminate based on such issues as skills, education, experience, ability and even personal preference so long as such discrimination is not also based on or related to race, color, national origin, sex, religion, age, mental or physical disability, or veteran status (Figure 7 – 6).

Discrimination, as described in the various employment laws, is against the law. In that sense, discrimination means all actions taken against individuals in the protected classes that have the effect of treating them disadvantageously and differently from other people not in the same category (Figure 7 – 7). Protected classes, again, are based on race, color, national origin, sex, religion, age, mental or physical disability and veteran status.

It's really not hard to know or sense when an employer is discriminating against employees in a manner that is against the law. It is often hard to prove it. That's why governments set up agencies such as the Equal Employment Opportunity Commission (EEOC) to enforce the intent and purpose of the law and provide assistance to employees who feel they have been the victims of discrimination. Most states and cities have similar organizations to assist in enforcing their equal employment opportunity laws.

Generally, these agencies and the courts use three major principles to determine discriminatory employment practices that may not be apparent on their face. First is the principle of discrimination by effect, even if not intended. Certain acts can constitute unlawful discrimination because of their effect on protected classes, even if not purposeful or intended.

For example, if a company has a policy of adhering to the EEO law and there are significant numbers of African-Americans in the natural hiring area of the company, and African-Americans did apply for various jobs with the company, and it turns out that over a period of time (usually years) the company did not hire any African-Americans, then there is a presumption that discrimination is occurring, whether the company intended it or not.

Second, the principle of unequal or disparate treatment has evolved from EEO law enforcement. Putting it simply, if similarly situated or equally qualified persons receive unequal treatment, discrimination has occurred. If the person who has suffered such discrimination is in a protected class, then the action was unlawful.

For example, assume that Jim and Jane are both lab technicians in a pharmaceutical company and have been employed for about the same length of time. They do very similar work, with a few minor differences in duties. Jim, however, is earning $1.50 per hour more than Jane. On its face, that's unequal or disparate treatment of a member of a protected class.

Third is the principle of unequal or disparate impact. When a practice or procedure may appear neutral on its face, but its application may fall more heavily on members of a protected class, it constitutes unequal or disparate impact.

For example, assume a company adopts a medical policy that does not pay benefits covering liver transplants, heart transplants or pregnancy (before the Pregnancy Discrimination Act of 1978). Men and women have livers and hearts, so that's equal. However, not many men get pregnant, so there is an unequal or disparate impact on a protected class. In fact, that's pretty much why the Pregnancy Discrimination Act of 1978 was adopted.

It is through the application of these three principles that EEO enforcement agencies determine whether the law is being broken in those cases where it is difficult to prove (Figure 7 – 8).

Stereotypes

One of the key things the law emphasizes is that employment decisions cannot be based on stereotypes or class assumptions. You must not assume certain kinds of people are better at some

Figure 7 – 8. Proving the Hard Cases

Three key principles used by EEO enforcement agencies to establish discrimination when it's not so obvious:

1. Discrimination by effect, even if not intended

Certain acts can constitute unlawful discrimination because of their effect on protected classes, even if the effect was not purposeful or intended.

2. Unequal or disparate treatment

If similarly situated or equally qualified persons receive unequal treatment, discrimination has occurred. If the suffering person is in a protected class, then the action was unlawful.

3. Unequal or disparate impact

When a practice or procedure may appear neutral on its face, but its application may fall more heavily on members of a protected class, it is unlawful.

things or don't like to do other things. For example, you must not assume that women would not be interested in an outside manual job (Women and men have both been doing that sort of work for centuries.), or that women aren't interested in long-range career paths (Given that most women work because they need to work, that's ridiculous.), or that women have a much higher turnover rate than men (It just isn't true.). Those are all stereotypes that have no foundation in fact. Thinking based on stereotypes will almost always get you in real trouble both in terms of the law and because they will often lead you to select the wrong people for a job. All employment-related decisions must be based on individual merit rather than stereotypes or class assumptions (Figure 7 – 9).

Figure 7 – 9. Get Rid of Stereotypes

All employment-related decisions must be based on individual merit rather than stereotypes or class assumptions.

Reasonable Accommodation

Problems relating to discrimination against the disabled or people practicing a particular religion almost always concern the issue of reasonable accommodation. Under the law, an employer must provide reasonable accommodation in these two cases. This means that the job environment and schedules must be tailored to make it possible for disabled or religious people to work where reasonably practical. This should be done without unreasonable hardship to the employer and/or fellow workers. Thus, Jewish employees should be allowed to celebrate Jewish holidays if Christian employees are allowed to celebrate Christian holidays (Figure 7 – 10).

The question, of course, is what is reasonable? It is what it seems it would be. For example, if your business happened to have its peak periods on Friday evenings and all day Saturday, and you required that everybody work these days, you could probably make a case that letting anyone off then is unreasonable. On the other hand, if any employees were allowed off Friday evening and Saturday, it would certainly be unreasonable for the employer to refuse to give that time off to a Jewish, Muslim or Seventh Day Adventist employee. For them, Friday evening and Saturday are religious days.

Issues pertaining to disabled workers are treated similarly. If a blind person applies for a job as a typist, and demonstrates that with special equipment he/she can perform the work, the employer should be able to reasonably accommodate this. The equipment may take a little more space, and the effort may take a little longer, but that is what reasonable accommodation implies. Employers are not required to accommodate in these instances if accommodation creates unreasonable costs or unreasonable hardship to the employer or fellow workers.

Figure 7 – 10. Reasonable Accommodation of Religion and Handicaps

Adjust the job environment and schedules to the needs of handicapped or religious workers where reasonably practical.

This should be done without unreasonable hardship to the employer and/or fellow workers.

Sexual Harassment

Applying the law with regard to sexual harassment sometimes gives some male Supervisors and managers particular problems. They seem to mix up personal feelings and needs with appropriate workplace behavior. Many Supervisors find it difficult to enforce appropriate behavior on this issue because of what they may see as custom and practice. The reality of today and tomorrow is that they had better learn fast because sexual harassment in the workplace is against the law. The inappropriate behavior must be corrected or the offending employee dismissed. In addition, the company and its agent (often a Supervisor as an individual) can be held liable for wrongs committed by the offender.

The EEOC has provided very clear guidelines on sex harassment:

Unwelcome sexual advances, requests for sexual favors and other verbal or physical conduct of a sexual nature constitute sexual harassment when (1) submission to such conduct is made a condition of employment, (2) submission to or rejection of such conduct is used as the basis for employment decisions or (3) such conduct has the purpose or effect of unreasonably interfering with an individual's work performance or creating an intimidating, hostile or offensive working environment.

The first two of these three guidelines are obvious, and most people understand them. The third guideline takes matters a very important step further. The question isn't whether the offender intended a sexual remark to be harassing; it is how the employee who was the object of the remark perceived it. The moral of that is clear: cut it out.

Age and Other Forms of Harassment

Complaints under the Age Discrimination in Employment Act (ADEA) are growing rapidly. This is because of the newness of the law, the changing workforce (more older people in and remaining in the workforce) and the growing awareness and assertiveness of older workers as they learn and use their rights under ADEA.

Under ADEA, an employer has an affirmative duty to maintain the work environment free from age harassment, including intimidation or insults. Illegal harassment may involve direct bias, such as a Supervisor telling jokes about older employees and stating that you can't teach an old dog new tricks. Other forms of harassment include assigning an older employee to less desirable work (*You've been around longer, so you know how to do this.*), calling someone *Pops*, making other negative references to the individual's age, such as, *I know it's probably hard for you to change these habits after all these years, but....* Just as with other forms of harassment, an employer is liable for harassment by co-workers as well as by Supervisors.

The following guidelines will help you avoid age discrimination:

- Disregard age in making all personnel decisions, whether it's hiring, promotion, termination or job duties and assignments.
- Review policy to be sure it does not discriminate against those over age 40, even if the policy is applied neutrally to all employees, unless you are prepared to prove it's a necessity on the job. (Example: High physical standards required for a job.)
- Avoid age-based criteria in hiring selection, such as young and attractive, youthful appearance or recent college graduate.

Figure 7 – 11. Rules for Avoiding Harassment Charges

- DO set a good example as a Supervisor.
- DO take complaints seriously, even if your first judgment is that the complaint is trivial or unwarranted.
- DO investigate complaints and take corrective action.
- DO know your company's policy on harassment of employees and communicate this to your staff.
- DO treat these matters confidentially and delicately.
- DO make sure that your employees know how to file a grievance or complaint and that they are encouraged to do so if harassment is occurring.
- DON'T use your position as Supervisor to request personal favors of any kind.
- DON'T wait for a complaint if you personally observe offensive behavior.

- Prepare a written explanation of why an older employee was rejected for promotion.
- If an older employee is terminated for poor performance or some factor other than age, have full documentation.

Employers must not retaliate against employees for filing a discrimination or other employment law complaint. If an employee feels discriminated against, he/she has the right to bring up the matter with the Supervisor, other persons in management or a government agency. Any retaliation for making a complaint is strictly illegal. Whether the retaliation is obvious (discharge) or subtle (denying a merit increase due to an employee's uncooperativeness), it is illegal. Complaints should not be viewed as a sign of disloyalty. They should be taken seriously and investigated objectively.

Questions about Harassment

Essentially, harassment is illegal. That means harassment related to race, sex, age, disability, etc. If you see an employee harassing another employee, confront the behavior and stop it. Check your own habits. For example, calling white employees by last names, but African-American employees by first names is a common form of harassment because it is demeaning, intentional or not (Figure 7 – 11).

Let's take a look at some of the questions you might have about harassment in the workplace. These questions are based on a list developed by the Equal Employment Advisory Council, an organization that assists employers in enforcing the Equal Employment Opportunity laws.

Q. What should I tell my employees about harassment?
A. Tell employees that harassment of any form (racial, sexual, national origin, etc.) is a violation of the company's policy on discrimination. Refer them to the letter on company bulletin boards or the company policy manual that spells out the implications of this policy. Pass on to

them the definition of harassment given in this chapter and offer to answer any questions they may have on the subject.

Q. What impact does harassment have on our workforce?

A. Aside from obvious legal liabilities, it reduces the effectiveness and productivity of the workforce. It may also increase turnover and cause an organization to lose good employees. When people are not accepted or feel intimidated by the majority members of a work group, they are not motivated toward the full development of their potential or commitment to the company.

Q. How does harassment differ from discrimination?

A. For all practical purposes, they are the same. Harassment is a form of discrimination and is an unlawful practice.

Q. When may male/female interpersonal relationships constitute harassment?

A. A number of actions can constitute harassment but a good rule to remember in this case is that harassment is present when one or the other individual indicates advances or attentions are unwanted and such advances or attentions continue. Touching, patting, poking—these can all be harassment. If someone says stop, then stop. Preferably, don't start in the first place.

Q. How do you handle situations where two employees are dating, going steady, etc.?

A. This has nothing to do with harassment. It is possible that employees working together may be romantically attracted to each other, and as long as this relationship does not interfere with their individual job performances, it is not a matter of company concern. If their relationship ends, and one employee keeps making unwelcome advances to another, then it is harassment.

Q. What do you do in a situation where an employee complains of harassment but only gives general information and will not reveal specific names or events?

A. Record the complaint. In absence of more complete details, conduct a low-key investigation to determine the possible basis for the complaint. Advise the employee of the actions taken. As a normal rule, such complaints should be viewed as an early warning signal since the offending behavior is likely to be repeated. An employee may not want to stir up trouble for another employee but if no changes occur, he/she may feel compelled to file a specific complaint. If there is any possibility that harassment is involved, one way to defuse this situation is to remind all employees in the work group of the company's policy.

Q. What is likely to happen if a Supervisor does not take the employee's complaint seriously?

A. The Supervisor places the company in jeopardy of being found in intentional violation of the law by an EEO enforcement agency. In such instances the enforcement agency could take the position that the company was placed on notice and intentionally failed to act. Supervisors should take any comments or statements seriously, no matter how casual, and report the incident to the company's EEO Officer.

Q. What do you do about a complaint of harassment that is occurring off the job?

A. As a general rule, the company should not become involved with the private lives of employees. However, Supervisors may be viewed as company representatives when off the job, depending upon the circumstances. If the complaint alleges harassment by a Supervisor or manager, you should report it, and the company should look into the matter. If the harassment is occurring between peer employees and does not involve a Supervisor or manager, the company normally should not become involved. In such cases, special attention should be paid to the working relationship of the involved employees to assure that there is no carry-over harassment at the work site. However, harassment between peers at a business lunch, traveling on company business or at a company event held off company property are areas that should concern you.

Q. What do you do when you receive a complaint from an employee whom you think is mad at another employee and is trying to get him/her in trouble?

A. Accept the complaint; do not make assumptions. Each complaint should be considered bona fide until you have the results of a company investigation.

Q. What do you do if you determine that an employee has made a false accusation?

A. Handle it like other employee disputes. Do not accuse the employee of lying unless you have undisputed evidence. The best way is to tell the employee that you were unable to substantiate the claim and can take no further actions until the employee can provide additional evidence.

Q. Should employees be permitted to use ethnic, racial or commonly used slurs where it has been a common practice to use these terms in a give-and-take manner?

A. Some employees may tolerate this practice because they don't want to rock the boat, but that doesn't mean that they might not be deeply offended by such language. This practice should not be permitted and is totally unacceptable.

Q. Are contractors' employees and other noncompany workers at the work site covered by the company's harassment guidelines?

A. Yes. They are expected to meet the same behavior standards as company employees.

Q. Can an employee file a lawsuit against a Supervisor or co-workers directly without first filing a charge of harassment with an EEO enforcement agency?

A. Yes. Individuals can be sued for personal liability under various federal and state statutes for actions such as assault, battery, emotional distress, etc.

Avoiding Employment Law Claims

As a Supervisor, one of your jobs is to maintain the workplace so that employment law claims do not arise (Figure 7 – 12). Difficult as it may sometimes seem, the steps to achieving it are straightforward.

Here are a few guidelines that will help you to avoid employment law claims. First, consider your state of mind.

- Make sure you apply all rules equally to everyone.
- Always send consistent signals and be honest in your appraisals.
- Avoid delaying decisions or ignoring problems.
- Assume everyone wants to advance in the job and with the company.
- Give clear instructions and warnings of consequences.
- Always hear the employee's side of the story before taking action.
- Avoid reaching conclusions on the basis of subjective feelings about employees; try to stick with objective facts.
- Clearly explain decisions to those employees affected.
- Keep those communication channels open!

Secondly, be aware that there are certain activities or times in which you need to be sensitive to discrimination issues. These include most of the major employment status actions you take, namely:

- Interviewing, hiring and orientation
- Carrying out a Performance Appraisal
- Making decisions about promotions
- Changing duties and assignments
- Revising work and employee schedules
- Changing working conditions
- Handling gripes, grievances and complaints
- Making decisions involving discipline and discharge

Figure 7 – 12. EEOC Claims against Employers

The following are a few examples of EEOC cases against employers that the employers lost:

- Using unlawful hiring criteria adversely affecting women and minorities, such as arrest records, height, weight and family status.
- Imposing an education requirement without showing its job relatedness.
- Relegating women and minorities to low-paid and undesirable jobs.
- Excluding minorities from supervisory, sales, skilled and technical jobs.
- Laying off female employees while retaining males with less seniority.
- Limiting females from overtime work.
- Using stiffer promotion criteria for women/minorities.

- Settling disputes between yourself and an employee or between employees

Employment law claims or charges of discrimination arise when an employee feels unfairly or insensitively treated. Understanding that you are supervising in a democratic workplace is a first step toward sensitivity. The second step is to realize that for most minority and protected-class workers, the workplace is still dominated by white males. Even if you, their Supervisor, might be a member of a protected class, it is probable that most of the other Supervisors and managers they see are not from a protected class. So you need to be sensitive to the reality of the workplace from their point of view.

The best way to express that sensitivity is precisely what we have been describing in this chapter and other chapters. As the Supervisor, you must engage in honest and fair dealing, clear and candid communication and objective evaluation of individuals and situations. Above all, show that you have an ongoing regard for every employee's personal dignity and worth. That will help avoid complaints and make for a good Supervisor.

Disciplining a Protected Class Employee

The subject of disciplining employees will be covered in detail in Chapter 11, Discipline in the Workplace, where we stress the need for using the progressive discipline approach. Proper discipline procedure is the same for all employees, protected or not. In the present context of giving you some aids to avoid facing employment law charges or claims, we will summarize the basic principles involved in effective disciplinary action for employees with performance problems. These apply to protected classes and all other employees.

- Clearly communicate your expectations.
- Try to rehabilitate the employee through training, coaching and detailed performance expectations.
- Document your efforts and the employee's response.
- Apply equal treatment (same standards and expectations) to all employees.
- Discharge, if necessary, should be the final step of a progressive disciplinary process (which can include steps like warnings, probation, suspension).

In summary, there should be no basic difference in how you discipline those in protected classes and how you discipline other employees. Principles of sound, equitable and considerate supervision should apply to everyone. The starting point of good employee relations is recognition of each person's unique individuality and the conviction that he or she will respond most favorably when treated with respect and thoughtfulness.

Affirmative Action

The United States policy on Affirmative Action is mandated by Presidential Executive Order (E.O.) 11246, instituted in 1965. No President since 1965 has chosen to eliminate or markedly alter E.O. 11246. The courts have also upheld E.O. 11246. It is the official policy of the United States.

The purpose of Affirmative Action is to help accelerate participation of the protected classes in the workforce. It is meant to be a proactive measure rather than a reactive measure. In other words, it seeks to undo the wrongs of the past and present by adopting policies that will affirmatively advance members of the protected classes in today's and tomorrow's workplace. Recent Supreme Court decisions, often described as attacking Affirmative Action, have, in fact, sustained it while at the same time asserting that those who might be adversely affected by Affirmative Action must also be taken into consideration.

Most companies are required to have Affirmative Action plans. Such plans usually require that areas of underutilization of protected classes be identified. Such areas can be identified by comparing the numbers of protected class workers to non-protected class workers in the local regional hiring area and within the various jobs in the company. A plan is drawn up to address the problem and seek an appropriate balance among employees. Such plans call for reasonable and reachable goals (not quotas), with methods and timeframes for achieving them but without reverse discrimination (defined in the next section). The Affirmative Action plan does not provide for the lowering of performance or conduct standards.

Affirmative Action plans should also include reasonable accommodation for the disabled. For example, at least some washroom stalls should be wide enough for a wheelchair, and telephones with amplifiers should be provided for the hearing impaired.

As a Supervisor, you should familiarize yourself with your company's Affirmative Action plan and do your part in seeing that the goals are met. The thrust of these employment laws is to extend the rights of reasonable treatment. The idea is that employers must act in a fair manner, and co-workers must act similarly. However, that concept is an evolving one, and promoting Affirmative Action for protected classes has sometimes resulted in charges of reverse discrimination.

Reverse Discrimination

Reverse discrimination is a popular (rather than legal) concept. For all practical purposes, reverse discrimination refers to the negative impact of Affirmative Action plans on nonprotected class white males. In other words, reverse discrimination could probably be subtitled Affirmative Action and the White Male Who Isn't an Ethnic, Over 40, a Veteran or Disabled. The point is, where does protected class and Affirmative Action leave the white male worker, as defined above? Frankly, it is still unclear.

For example, consider the following theoretical situations.

- A white male has a cause of action based on sex discrimination if he works in a female-dominated company and was denied promotion repeatedly so that females might have the position.
- A white male has a cause of action based on race dis-

crimination if he works in a minority-dominated company and was denied promotion repeatedly so that minorities might have the position.

However, problems arise because the workplace is still largely dominated by white males. The law recognizes that and seeks to redress the disproportionately lower share of jobs held by members of the protected classes. Of course, if you are a white male and you do not exactly dominate your workplace, none of this makes much sense to you personally. In fact, you may feel absolutely discriminated against if your company has a very active Affirmative Action policy and you find people being affirmatively promoted over you.

This is what is often called reverse discrimination. The confusion over this concept arises because it is not written in the law (except for sex discrimination). It is a judicial concept that is only slowly evolving.

For example, consider these current principles under our employment laws evolving out of situations in which white males are the dominant group in a workplace.

- It is very clear that sexual harassment of a white male is just as much against the law as sexual harassment of a female.
- It is very clear that an Affirmative Action policy that affirmatively awards jobs to minorities and women over equally qualified white males is within the law.
- It is probably true that an Affirmative Action policy that affirmatively awards jobs to minorities and women over marginally better qualified white males is within the law.
- It is possible that an Affirmative Action policy that affirmatively awards jobs to minorities and women who are not qualified for the job over clearly qualified white males is against the law. However, there really has not yet been a case that clearly establishes this in the workplace.

Recent Supreme Court decisions are beginning to more clearly define the rights of those who are adversely impacted by Affirmative Action plans, but their thrust does not appear to fit the image of reverse discrimination. Instead, it simply points to the need to protect one group from adverse impact while, at the same time, proceeding with Affirmative Action. Like many other aspects of employment law, reverse discrimination is a long way from being a settled issue.

WORKPLACE LAW

Workplace law seeks to assure that employees work in a safe workplace, that they are paid a decent wage for their labor, and that they are assured certain basic conditions of work and rights to representation. There are many workplace laws. We have tried to select those laws that will impact Supervisors the most. We have not gone into great detail on safety laws, since the National Safety Council has other books and courses that cover these subjects in great detail. (See, for example, *Accident Prevention Manual for Industrial Operations,* latest edition and *Fundamentals of Industrial Hygiene,* latest edition.)

The OSHAct

The Williams-Steiger Occupational Safety and Health Act of 1970, generally referred to as the OSHAct, became effective April, 1971, with the creation of OSHA (Occupational Safety and Health Administration). The OSHAct seeks to establish rights to a safe workplace, determine what is safe and establish an enforcement procedure to see that safe standards are followed.

Each employer has clear obligations under the Act to do the following:

- Furnish to each employee both employment and a place of employment that are free from recognized hazards causing or likely to cause death or serious harm to employees.
- Comply with occupational safety and health standards promulgated by the Act.

As a Supervisor in charge of your part of the workplace, these employer obligations become your obligations! A Supervisor is directly responsible for the safety of employees (Figure 7 – 13).

OSHA's Mission

As OSHA's mission has developed, the emphasis has varied. Stringent enforcement of rules and guidance in meeting rules have changed with national leadership. As we have become more accustomed to OSHA, some of the first educational processes both for workers and management may seem in some cases minimal and in other cases excessive. As OSHA leadership found a need to stress one area of enforcement over another, many viewed the changes as confusing. OSHA's mission, on the other hand, is very straightforward, and few would object to what it provides. Essentially, Congress directed OSHA to:

- Encourage employers and employees to reduce workplace hazards and implement new or improve existing safety and health programs.
- Conduct and provide for research in occupational safety and health and develop innovative ways of dealing with occupational safety and health problems.
- Establish separate but dependent responsibilities and rights for employers and employees for the achievement of better safety and health conditions.
- Maintain a reporting and recordkeeping system to monitor job-related injuries and illnesses.
- Develop mandatory job safety and health standards.
- Provide for the development, analysis, evaluation and approval of the state occupational safety and health programs.

While a lessor known function, OSHA also monitors maintenance of the general sanitation and working conditions of the

Figure 7 – 13. OSHAct of 1970

OSHA says employers must:

- Furnish to each employee both employment and a place of employment that are free from recognized hazards causing or likely to cause death or serious harm to employees.
- Comply with occupational safety and health standards established by OSHA.

Figure 7 – 14. OSHA'S Five Categories of Violations

1. Imminent Danger
Could be expected to cause immediate death or serious physical harm.

2. Serious
Likely that death or serious physical harm could result.

3. Willful and Repeated
Employer knows about violation. Employer keeps doing it.

4. Nonserious
Violation that is not serious but does have a direct impact on occupational safety and health.

5. De minimis
Minimal violation with no immediate or direct impact on safety and health of the employees.

Figure 7 – 15. Supervisor's Checklist of Training Subjects on Occupational Safety and Health

- Proper observance of safety regulations
- Routing for emergency exit
- Accident and injury treatment
- Occupational health and environmental controls
- Hazardous material
- Personal protective equipment
- Medical and first aid
- Fire protection
- Materials handling and storage
- Machine guarding
- For welding, should cover cutting and brazing

workplace. This means OSHA must also look at things like housekeeping, extermination and efforts to maintain a healthy and reasonably clean workplace.

Five Categories of Violations

Congress determined and OSHA established the details on five categories of safety/health violations or potential hazards. You need to be aware of these, and the level of importance each has.

1. **Imminent Danger.** This covers any conditions or practices that could be expected to cause immediate death or serious physical harm. In such circumstances, the company can be shut down under OSHA regulations.
2. **Serious.** These are conditions where there is a substantial probability that death or serious physical harm could result from the alleged violation. The company may be fined whether the employer knew of or, with reasonable diligence, should have known of the hazard.
3. **Willful and Repeated.** Violations are considered willful when the employer either intentionally or knowingly violates the Act or, although not consciously violating it, is aware of hazardous conditions without making an effort to eliminate them. Repeated violations are those for which OSHA has issued a second citation for the same act.
4. **Nonserious.** This is a violation that is not serious but has a direct or immediate relationship to occupational safety and health.
5. **De minimis.** De minimis means a minimal violation that has no immediate or direct relationship to the safety and health of the employees (Figure 7 – 14).

Accident Reporting

OSHA calls for specific recordkeeping of accidents that occur at the workplace or while performing work-related activities. As a Supervisor, you may have the responsibility of initiating or maintaining these records. You are under a legal obligation to maintain accurate, honest and clear records. Inaccurate recordkeeping is bad policy for everyone and can lead to significant fines and penalties.

OSHA-Required Training

Under the regulations, employers are expected to train new employees and to have an ongoing training program on issues of workplace safety and health. This training is most often carried out by Supervisors. The National Safety Council also offers many of the required training programs in its Safety Training Institute.

Training should provide for proper instruction and observance of safety regulations, routing and procedures for emergency evacuation, and accident and injury treatment (Figure 7 – 15). Training should also specifically cover hazard communication (chemical exposure), occupational health and environmental controls, hazardous material, use and care of personal protective equipment, medical and first aid, fire protection, materials handling and storage, machine guarding, and cutting and brazing for welding operations.

Employee Rights under OSHA

OSHA has provided employees with a Bill of Rights when it comes to maintaining occupational safety and health (Figure 7 – 16). You need to know these rights, since the assertion of the rights by an employee will most often first be directed to you, as the Supervisor.

Employees have a right to:

- Request an OSHA inspection if they believe an imminent danger exists or that a violation of a standard exists that threatens physical harm
- Have a representative accompany an OSHA compliance officer during the inspection of a workplace
- Advise an OSHA compliance officer of any violation of the Act that they believe exists in the workplace and privately question and be questioned by the compliance officer
- Have regulations posted by the employer to inform them of protection afforded by the Act
- Have locations monitored by the employer in order to measure exposure to toxic or radiation materials, have access to the records of such monitoring or measuring and have a record of their own personal exposure
- Have medical examinations or other tests to determine whether their health is being affected by an exposure and have the results of such tests furnished to their physicians
- Have posted on the premises any citations made to the employer by OSHA.

As you can see, this is a very extensive and clear-cut list of rights. No disciplinary action may be taken against an employee for asserting these rights. In fact, we strongly recommend that you and your employer insist on and maintain complete compliance with these OSHA standards for a safe and healthy

Figure 7 – 16. Employee Bill of Rights under OSHA

Employees have a right to:

- Request an OSHA inspection
- Have a representative accompany OSHA compliance officer
- Privately advise and question OSHA compliance officer
- Have OSHA regulations posted in workplace
- Have toxic or radiation materials exposure monitored
- Have access to the records of such monitoring
- Have a record of their own personal exposure to material
- Have medical examinations or tests to determine if health is affected by exposure, with results of tests furnished to their physicians
- Have posted on the premises any citations made to the employer by OSHA

workplace. They make sense. They are right. They improve productivity.

OSHA Self-Inspection Checklist

While your employer's safety officer will probably have primary responsibility to ensure these standards are met, you as a Supervisor are in charge of the day-to-day implementation and application of the standards. Therefore, you should regularly check your work area for these factors (Figure 7 – 17).

OSHA Visitations

It is useful to know what an OSHA compliance officer will be reviewing. Compliance officers start by checking recordkeeping. Based upon the records, they may look for the following hazardous conditions:

- Eye irritation
- Strong odors
- Visible dust or fumes in the air
- Excessive noise
- Spilled or leaking chemicals
- Use of substances known to be dangerous even when handled properly

In addition, officers will check:

- Changes in the physical work environment or work procedures made by the employer to reduce health hazards
- Effectiveness of personal protective gear
- Maintenance of protective gear
- Training in proper use of protective gear
- Isolation of eating, washing and resting areas from work areas containing hazardous substances

HazCom

The OSHA Hazard Communication Standard now extends to all employers. This new regulation is being dealt with separately because it expands the rights of employees and the obligation of employers under the Occupational Safety and Health Act. In general, the Hazard Communication Standard pre-empts any state statute of a similar nature.

Essentially, the standard has one purpose: the assertion of people's right to know about hazardous chemicals they may be exposed to at work (Figure 7 – 18). The standard requires that the employer carry out the following steps in order to implement this very simple right:

1. Develop a written communication plan that explains the organization's hazard program.
2. Provide a system to ensure chemicals are properly labeled.
3. Conduct a chemical inventory in order to establish the possible chemicals that individually or in combination provide a potential hazard.

4. Obtain material safety data sheets (MSDS's) from chemical suppliers and make them available to employees.

5. Establish an employee training program that covers:
  * Physical health hazards of the chemicals
  * Methods and observations that may be used to detect the presence or release of hazardous chemicals in the work area
  * Protective measures available (i.e., equipment, work practices, etc.)
  * Retraining when a new hazard appears, that is, other chemicals, procedures, new techniques for processing, etc. (Figure 7 – 19)

Fair Labor Standards Act, 1938 (FLSA)

The FLSA asserted the right of Congress to regulate the minimum standards for hours and wages of employees covered by the Act, child labor and overtime (Figure 7 – 20). Among other reasons, it came into being because workers were required to work long hours at low wages and were often so tired that they created major safety hazards for themselves and others. Today, the standards set under FLSA are the bedrock for most of our workplace law.

Minimum Wage

Most industries and employers are regulated under FLSA. Since minimum wage is set by Congressional Act, it doesn't change often. The longer the time between changes, the fewer people who actually benefit under the minimum provisions. For instance, before it was changed in 1989, the minimum wage was $3.35/hour and very few workers were receiving a wage that low. In certain industries, notably hotels and restaurants, there are some exceptions to the minimum wage.

Child Labor

FLSA also regulates child labor. Currently, the law provides that people 18 years and above can hold any job, hazardous or not with unlimited hours, only children 16 – 17 years old can hold any nonhazardous job with unlimited hours, and children under the age of 16 have numerous restrictions, such as hours, conditions and parental approval.

Overtime

FLSA also regulates overtime rates for employees who are nonexempt. Nonexempt, in a sort of logically reversed meaning of the word, means employees who are covered by special regulations of FLSA. Under these regulations, overtime is defined as over 40 hours in a workweek. A workweek is seven consecutive days. Thus, these hours may not be averaged over a two- or three-week period, for instance. Overtime pay must be at least 1.5 times the regular rate. Regular rate is defined as the rate per hour actually paid the employee for the normal, nonovertime workweek. As a Supervisor, you often have responsibility for filling out the timesheets, so this requirement must be applied for your nonexempt employees.

Figure 7 – 17. OSHA Self-Inspection Checklist

1. Is the required OSHA workplace poster displayed in your place of business as required where all employees are likely to see it?
   OK _____ Action Needed: _____

2. Are you aware of the requirement to report all workplace fatalities and any serious accident (where 5 or more are hospitalized) to a federal or state OSHA office within 48 hours?
   OK _____ Action Needed: _____

3. Are workplace injury and illness records being kept as required by OSHA?
   OK _____ Action Needed: _____

4. Are you aware that the OSHA annual summary of workplace injuries and illnesses must be posted by February 1 and must remain posted until March 1?
   OK _____ Action Needed: _____

5. Are you aware that employers with 10 or fewer employees are exempt from the OSHA recordkeeping requirements, unless they are part of an official Bureau of Labor Standards or state survey and have received specific instructions to keep records?
   OK _____ Action Needed: _____

6. Do all employees know what to do in emergencies?
   OK _____ Action Needed: _____

7. Are emergency telephone numbers posted?
   OK _____ Action Needed: _____

8. Are all electrical cords strung so they do not hang on pipes, nails, hooks, etc?
   OK _____ Action Needed: _____

9. Is there no evidence of fraying on any electrical cords?
   OK _____ Action Needed: _____

10. Are metallic cable and conduit systems properly grounded?
    OK _____ Action Needed: _____

11. Are portable electrical tools and appliances grounded or double insulated?
    OK _____ Action Needed: _____

12. Are switches mounted in clean, tightly closed boxes?
    OK _____ Action Needed: _____

13. Are all exits visible and unobstructed?
    OK _____ Action Needed: _____

14. Are all exits marked with a readily visible sign that is properly illuminated?
    OK _____ Action Needed: _____

15. Are there sufficient exits to ensure prompt escape in case of emergency?
    OK _____ Action Needed: _____

Figure 7 – 17 Continued

16. Are portable fire extinguishers provided in adequate number and type?
    OK _____ Action Needed: _____

17. Are fire extinguishers recharged regularly and properly noted on the inspection tag?
    OK _____ Action Needed: _____

18. Are fire extinguishers mounted in readily accessible locations?
    OK _____ Action Needed: _____

19. Are NO SMOKING signs prominently posted in areas containing combustibles and flammables?
    OK _____ Action Needed: _____

20. Are waste receptacles provided and are they emptied regularly?
    OK _____ Action Needed: _____

21. Are stairways in good condition with standard railing provided for every flight having four or more risers?
    OK _____ Action Needed: _____

22. Are portable work ladders and metal ladders adequate for their purpose, in good condition and provided with secure footing?
    OK _____ Action Needed: _____

23. Are all machines or operations that expose operators or other employees to rotating parts, pinch points, flying chips, particles or sparks adequately guarded?
    OK _____ Action Needed: _____

24. Are mechanical power transmission belts and pinch points guarded?
    OK _____ Action Needed: _____

25. Are hand tools and other equipment regularly inspected for safe condition?
    OK _____ Action Needed: _____

26. Are approved safety cans or other acceptable containers used for handling and dispensing flammable liquids?
    OK _____ Action Needed: _____

27. Are your first-aid supplies adequate for the type of potential injuries in your workplace?
    OK _____ Action Needed: _____

28. Are hard hats provided and worn where any danger of falling objects exists?
    OK _____ Action Needed: _____

29. Are protective goggles or glasses provided and worn where there is any danger of flying particles or splashing of corrosive materials?
    OK _____ Action Needed: _____

Figure 7 – 18. OSHA Hazard Communication Standard

People have a right to know about hazardous chemicals they may be exposed to at work.

Figure 7 – 19. OSHA Hazard Communication Standard

Employers must provide the following:

1. Written Communication Plan
2. Labels and Warnings
3. Chemical Inventory
4. Material Safety Data Sheets
5. Employee Training

Exempt Employees

Exempt employees (those employees not covered by these special regulations) are defined as:

- Administrative, executive and professional
- Commissioned salespersons of retail or service establishments
- Domestic service workers residing in the employer's residence
- Farm workers
- Railroad and airline workers
- Auto, truck, trailer, farm implement, boat or aircraft salespersons, partspersons and mechanics employed by nonmanufacturing dealers
- Outside salespersons
- Retail-service establishment (laundry-drycleaning businesses not included)

However, the actual determination of exemption is also dependent upon one's duties, not just upon job title. Generally, many Supervisors are exempt employees. However, the distinction depends on the duties and responsibilities assigned. While the minimum wage and overtime regulations do not cover these employees, there is a requirement that those in managerial positions be paid on a salary basis. This means that they are to be paid for any week they work, regardless of hours or days worked. FLSA provides specific tests to determine which positions fall into which category (exempt or nonexempt), but are too lengthy and cumbersome for our purposes.

The Equal Pay Act, 1963

The Equal Pay Act is an amendment to the FLSA. It prohibits unequal wages for women and men working in the same establishment, doing equal work that requires equal skill, effort and responsibility and that is performed under similar working conditions. The Act provides that the equal pay standard does not rely upon job classifications or titles, but depends on the actual job requirements and duties performed. It also provides that equal pay includes vacation and holiday pay, premium payments of any kind and fringe benefits.

There are three specific exceptions to this Act (Figure 7 – 21). Equal pay is not operable if the differential in pay is the result of a wage payment made (1) under a seniority system, (2) a merit system or (3) a system measuring earnings by quantity or quality of production.

Another, broader exception involves pay differential based on anything other than sex. It is important that you do not confuse equal pay for equal work with comparable pay for comparable work. The latter is a right that is being asserted but is not yet the law. The former is the law. Let's look at two examples to see the difference.

Example 1. Jim and Jill work at Compatible Computers. Jim is a Technician 1 and earns $16.50 per hour. Jill is a Technician

2 and earns $15.50 per hour. Both needed a college degree for the job. While their titles are different and their pay is different, an analysis of their duties shows that they perform substantially the same job. Jill, on that basis, is entitled to the higher rate of pay.

Example 2. Jane and John also work at Compatible Computers. Jane is a Technician 2 and earns $16.50 per hour. Jim is a Programmer and earns $22.75 per hour. Both needed a college degree for the job. Their titles are different and their pay is different, although they both had to have some of the same entry level requirements (college degree, in this case). However, a technician's work is significantly different from a programmer's. Technicians primarily work on operating and maintaining existing equipment, while programmers are developing new systems and procedures. The latter requires a greater depth of knowledge and experience. Therefore, there may be comparability but the work is not equal. Jane has no claim under current case law.

National Labor Relations Act (NLRA)

The NLRA is the basic labor law of the United States. In a nutshell, the NLRA establishes and protects the rights of employees to organize themselves into a labor union and to be represented by such a union to their employers. Representation usually comes about through a representation election conducted by the National Labor Relations Board (NLRB), the agency that administers the Act. The employer is not allowed to interfere in such organizing. Although entitled to express opinions about the organizing, the employer must do so without threat of intimidation or reprisal to the employees. After the union is certified by the NLRB, the union has the right to represent such employees for the purposes of negotiating with the employer over the terms and conditions of employment. These include wages, hours, benefits and conditions of work. The employer and the union must both bargain in good faith over these issues. They are not required to reach an agreement, although the thrust of the law is in that direction.

Bargaining Unit. Union elections involve only those employees included in a bargaining unit. The bargaining unit, generally, is a particular group of workers within a plant or facility who have a common interest. In an auto plant, this might be the production workers but might exclude the clerical workers. Supervisors are excluded from the election and the bargaining unit, except in the airline and railroad industries, which are covered by a different law (the Railway and Airline Labor Act).

Union Organizing Activity. If the facility where you work is nonunion and a union is trying to organize the workers, or if there is an existing union and a competing union is trying to take over, you, as a Supervisor, will inevitably be involved in this process as a representative of the company. If a union campaign or election is occurring in your workplace, you should familiarize yourself with the company's position. If that position is to resist the union, permitted activities include:

- Representing your company in a positive manner

Figure 7 – 20. Fair Labor Standards Act, 1938 (FLSA)

Regulates:

- The minimum standards for hours and wages of employees covered by the Act
- Child Labor
- Overtime

Figure 7 – 21. The Equal Pay Act, 1963

EQUAL PAY FOR EQUAL WORK

Exceptions:

- Seniority systems
- Merit systems
- Production/piecework systems

- Raising questions about employees' representation under a union contract
- Handing out employer literature

Remember that you are a representative of management. Management interference is prohibited during a union election. Therefore, you must avoid any actions that affect an employee's pay or job, arguments over questions of the union, threatening through a third party or dealing with any of the union's organizing officers without the advice of top management.

Specifically, you are not permitted to:

- Promise rewards for not joining the union
- Make threats about what might happen if the union represents employees
- Pressure employees to commit themselves to the company
- Ask if they've signed a representation card
- Spy on employee/union activity
- Invite employees into your office to privately discuss the union
- Discriminate against employees due to their union views or activities. Discrimination is defined as any action, such as discharge, layoff, demotion or assignment to a more difficult or disagreeable job on account of the employee's union stance or membership (Figure 7 – 22).

The Supervisor and the Union Contract

While we have previously dealt with this subject and will deal with it once again in Chapter 8, it doesn't hurt to reemphasize the very important role of the Supervisor in the administration of a labor agreement (Figure 7 – 23). If your company already has a union contract, you must familiarize yourself with it and know the requirements regarding assignments, pay, hours of work, duties, the grievance procedure, and so on.

Figure 7 – 22. The National Labor Relations Act and Union Organizing Activity

As a Supervisor you can:

- Represent your company in a positive manner
- Raise questions about employees' representation under union representation
- Hand out employer literature

As a Supervisor you cannot:

- Promise rewards for not joining the union
- Make threats about what might happen if the union represents employees
- Pressure employees to commit themselves to the company
- Ask if they've signed a representation card
- Spy on employee/union activity
- Invite employees into your office to privately discuss the union
- Discriminate against an employee for his/her union views or activities

Remember that the shop steward represents the employees on these issues, just as you represent the company. On many other aspects of work, you act in the role of agent for your employees, so it is sometimes difficult to keep things straight! In the context of the labor agreement and the items covered in it, you do not represent your employees. The shop steward does.

You should strive to develop a good understanding between yourself and the committeeperson, steward or business agent, always recognizing the differences, but always seeking harmony as a goal. To avoid problems, keep them informed about what's going on. Ignorance and surprises cause problems. In a very real sense, you both share one common task: the fair and proper application of the terms of the labor agreement. Seek their cooperation, but remember that you're the boss and this is not a co-supervisory relationship.

It is also useful to recognize that the union can help you by supporting the company safety program. Union involvement in safety committees can only reinforce the idea that safety is of universal importance.

CONCLUSION

While there are dozens of federal laws impacting on employment and the workplace and literally hundreds of state and local laws, we have briefly dealt with the ones that are likely to most directly affect your work as a Supervisor:

- Title VII of the Civil Rights Act, 1964
- Pregnancy Discrimination Act, 1978 (amendment to Title VII)
- Civil Rights Act, 1964 (EEOC)
- Age Discrimination in Employment Act, 1967 (ADEA)
- Rehabilitation Act, 1983
- State laws dealing with discrimination in the workplace
- OSHAct, 1970
- OSHA HazCom
- Fair Labor Standards Act, 1938 (FLSA)
- Equal Pay Act, 1963 (an amendment to the FLSA)
- National Labor Relations Act (NLRA)

Just reviewing so many laws is difficult enough. However, we also discussed the very difficult issues you must face:

- Assuring that you comply with employment and workplace law
- Assuring that your employees comply with them
- Assuring that you send a consistent message that the law is the law and must be respected

This chapter emphasized the following ideas:

1. The fundamental principles of employment and workplace law try to address the democratic rights of people and employees to fair, appropriate and safe treatment.

Figure 7 – 23. The Supervisor and the Union Contract

KEY IDEAS:

- The shop steward represents the employees' labor agreement issues; you represent the company.

- Develop a good understanding between yourself and the steward or business agent.

- Keep them informed about what's going on. Ignorance and surprises cause problems.

- You both share one common task: the fair and proper application of the terms of the labor agreement.

- Seek their cooperation, but remember that you're the boss and this is not a co-supervisory operation.

2. These laws evolve over time and are written in response to the needs of people and employees.
3. The laws not only must, but should, be enforced.
4. Part of your job is to see to it that this happens and that the people who work under your supervision understand why it is so.

Chapter 8

Supervision in a Union Environment

OVERVIEW

Supervising a unionized workforce has special implications for Supervisors and employees. In a union or nonunion environment, Supervisors have to be familiar with and enforce all the usual workplace rules, policies, procedures and workplace laws (Title VII, Fair Labor Standards Act, etc.). In a union environment, however, the Supervisor must also comply with and enforce a collective bargaining agreement or contract. When referring to what is often called a union agreement, union contract or labor agreement, we will use the appropriate term: *collective agreement*, or simply *agreement*. The term *collective agreement* is the most common term used in the law and industrial relations for two reasons: (1) It emphasizes the fact that the agreement was reached by labor and management, and (2) By agreement, it allocates responsibilities between them.

In this chapter, we will discuss who is involved in the collective agreement, the contents of a collective agreement and a company policy/procedure manual, legal considerations in a union environment and implications for the Supervisor in administering a collective agreement.

THE COLLECTIVE AGREEMENT

Parties to the Agreement

One of the easiest things for a Supervisor to forget is that the collective agreement is a result of collective bargaining. The parties to that bargaining, and to the collective agreement that results from it, are the union and the company. It is a contract between these two, and one that both have agreed to uphold and enforce (Figure 8 – 1). In that sense, the collective agreement is no different from any other contract the company makes. It was entered into by both parties in good faith, and both are obligated to live up to its terms. Thus, if you feel frustrated by the terms of a collective agreement, remember that your company signed and agreed to those terms.

Management in smaller companies often feels they had little choice in bargaining or accepting the collective agreement, since they often are involved in standby agreements. Standby agreements simply provide that the company agrees to accept whatever comes out of negotiations between the union and the industry. However, the company did sign the standby agreement, and thus

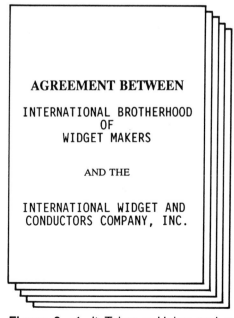

AGREEMENT BETWEEN

INTERNATIONAL BROTHERHOOD
OF
WIDGET MAKERS

AND THE

INTERNATIONAL WIDGET AND
CONDUCTORS COMPANY, INC.

Figure 8 – 1. It Takes a Union and a Company to Sign

113

agreed to sign the product of those negotiations. Any way you look at it, the company management chose to agree. They may have felt they had little choice, but is a collective agreement different from the contract signed when a major buyer of the company's product insists on better terms and conditions in its contract? How is the limitation of choice any different?

Finally, remember that the United States guarantees employees the right to be represented by a union. This guarantee is spelled out in the National Labor Relations Act, the Railway Labor Act, and in the laws of virtually all states (Figure 8 – 2). Those laws also provide that the company and the union have a legal obligation to bargain in good faith over the terms and conditions to be included in a collective agreement. If a majority of the employees truly want to be represented by a union, the company does not have a right to prevent this. Go back and review Chapter 7, Supervisors and the Law: Employment and Workplace Law, if you have any questions on this issue.

Topics Included in the Agreement

Generally, a collective agreement spells out the rights of the unionized employees, specifically those related to wages, benefits and working conditions. It also establishes a framework for the relationship between management (the Supervisor and those above the Supervisor) and the employees (those who are covered under the collective agreement). The question of who is or is not covered is determined by two issues: the law and the agreement between the union and management. Most people in management are automatically exempt, that is, not eligible to be covered by a collective agreement. This includes most Supervisors and foremen. In addition, the company and the union may agree to exclude certain employees from coverage. For example, in a factory it is often common to find clerical employees excluded from coverage. Sometimes this happens because they were not included in the unit involved in the union representation election, but sometimes they were simply excluded by mutual agreement between the company and union.

Virtually all collective agreements cover the same range of topics. How they deal with the topics varies significantly from industry to industry and even company to company. Typical topics are discussed in the following sections.

Parties to the agreement. *This agreement is by and between the AMALGAMATED MAINTENANCE, INC. and the INTERNATIONAL BROTHERHOOD OF COLLATORS, SCRIVENERS AND MAINTENANCE WORKERS OF AMERICA, AFL-CIO, CLC.*

A statement of purpose. *The purpose of this agreement is to improve and maintain the wages, benefits and working conditions of the employees of AMALGAMATED MAINTENANCE, INC.*

Recognition clause. *The Company recognizes the Union as the sole and exclusive bargaining agent for employees.* This means that the company will not deal with any other union for these employees until such time as a different union is selected by the employees. It also means that the union is obligated to represent

Figure 8 – 2. The National Labor Relations Act

The Company will not interfere with, restrain or coerce the employees covered by this Agreement because of membership in, or activity on behalf of, the Union.

all employees covered by the agreement even if they are not union members.

Coverage or scope. This clause usually spells out exactly which jobs and which locations are covered by the agreement, sometimes giving long lists of job titles and duties.

Union shop. Union shop rules provide that employees covered by the agreement must become members within a stated period of time after they are hired (usually 60 – 90 days). Some states prohibit such a clause. Some collective agreements provide that those who object to membership pay a maintenance fee in lieu of dues. Most unions see union shop rules as a critical part of their operations.

Hiring hall. This is most commonly found in the construction and maritime industry. The company agrees to hire employees from the union-operated hiring hall, with escape clauses against unqualified candidates.

Check-off. A check-off plan provides that union dues and fees will be deducted from wages by the company and paid to the union.

Discrimination. This clause usually prohibits two forms of discrimination: against union members and against individuals on the basis of race, creed, sex or religion. Both, of course, are covered under U.S. law, so the rule tends to be in the nature of a company proclaiming that it is an equal opportunity employer. Collective agreements have similar language, such as that shown in Figure 8 – 2: *The Company will not interfere with, restrain or coerce the employees covered by this Agreement because of membership in, or activity on behalf of, the Union.*

Hours of work. Assigned hours, tour of duty, hours per week, shifts, lunch hours and breaks are among the subjects covered here.

Overtime. This clause spells out when there is extra pay for working hours other than those regularly assigned.

Shift premium. Many agreements provide an extra payment or premium for working swing or midnight shifts. Twenty-four-hour-per-day operations such as railroads and airlines often do not provide such premiums.

Holidays. This clause usually spells out the precise paid holidays provided, determines eligibility and provides for extra premium pay if the employee is required to work on the holiday.

Call-in and reporting time. Call/report rules provide for how employees will be called in for special assignments, when they must report for duty and how they must report absences.

Vacation. Vacation rules spell out the precise vacation entitlement, usually based on length of service and actual days worked in the previous year. They determine eligibility and provide for an extra premium pay if the employee is required to work in lieu of taking an earned vacation.

Leave of absence. This covers the circumstances under which an employee is entitled to a leave of absence, usually including health, child bearing, military service and union duties.

Seniority. Considered the most critical rule by most unions and the most restrictive by many companies, seniority rules establish the principle that most job assignments shall be based on the employee's length of service with the company. This principle is usually modified by some reference to qualifications. Thus, it will often read: *In the assigning of positions, fitness and ability being sufficient, seniority shall prevail.* Note that the word *sufficient* is the critical word. Generally, arbitrators have placed a very narrow interpretation on this. Thus, sufficient does not mean *in comparison with others applying for the same job,* but simply *enough to start learning the new job.* A large number of grievances arise out of disputes in the assignment of jobs by seniority. Some companies, however, find assignment by seniority as good as any other system for production work.

Method of wage payment. This is a payday rule, establishing when and how frequently wages are paid.

Union representation, stewards and visitation. The agreement usually has a rule giving union representatives full access to visit the property, recognizing some rights for local shop stewards to take off time for union work. Some agreements provide that local shop stewards will be aided by the company while carrying out union duties. Most do not.

Grievance procedure. Grievances are disputes arising from the interpretation or application of the collective agreement. They occur during the life of the agreement and are usually settled through a company/union process of hearings and negotiations. Union agreements always provide for some method of grievance resolution. Usually it is a filing and appeal process, with the union having control over its representation of the employee. That is, the employee has the right to a union representative being present at each step of the grievance process, and the union has the obligation and duty to handle the grievances it finds appropriate and proper.

Arbitration. In the event the company and the union can't agree on how to settle a grievance, collective agreements usually provide that either party can choose to submit the dispute to a neutral third party, called an *arbitrator.* They also agree that the arbitrator's decision will be final and binding on both parties. This final step in the process assures closure of a dispute if the company and the union are unable to do so. Most unions and companies prefer to settle their own disputes. They tend to avoid arbitrations because (1) the process is expensive, and (2) they are turning their fate over to someone else. Some industries or companies have special arbitrators or arbitration panels that have been preselected by agreement between the union and the company. In these cases, the final step of arbitration is often viewed as less of a threat to self-interest, probably because the parties are familiar with the arbitrators.

Discharges. Discharge rules spell out the legitimate causes of discharges, how an employee is to be discharged and the rights of appeal involved. They often have some particular circumstances that can result in immediate dismissal. For example, in the banking and insurance industry, if an employee conceals prior criminal activity, the employee may be dismissed immediately. In the railroad and airline industry, employees drinking alcohol on the job are subject to immediate dismissal.

Bulletin boards. Collective agreements provide specific locations where union information can be posted, where company job notices and changes must be posted, etc.

Jury duty. Most agreements have clauses providing that employees have the right to be paid while on jury duty.

New jobs. Creation of new jobs is a subject for negotiation under most labor agreements. Posting and the method of filling such jobs is also usually spelled out.

Strike and lockout. The right to strike and lockout is provided by law. Most collective agreements provide that there will be no strikes or lockouts during the term of the agreement. The law does not require this, but most companies insist on the clause as an assurance of stability. In the railroad and airline industries (both are covered under a separate law, the Railway Labor Act), collective agreements do not expire. To get around this, companies have insisted that a clause be signed establishing a fixed period of time during which there will be no strikes or lockouts.

Waiver and alteration of agreement. This clause provides for exceptions to the agreement and how the agreement may be modified, usually a notice provision, need for a meeting to bargain collectively, etc.

Safety and health. This clause usually places clear obligation on the employer to maintain a safe work environment and on employees to follow safety rules and procedures. It often includes provisions for joint labor/management safety committees, assignments to committees, committee meetings and other steps to be taken to assure safety and health.

New employees. This spells out the method of hiring and training new employees, special pay circumstances and on-the-job training.

Death in the family. This most commonly provides paid leave for attending funerals for a death in the immediate family, including grandparents and close relatives, and sometimes bereavement time for death in the employee's immediate family.

Health and welfare. This commonly provides health, accident and life insurance programs for all employees covered by the agreement. It also may provide for a pension program in addition to Social Security. The terms of the coverage, carrier(s) for the plan(s) and administration of the plan(s) are usually covered in bargaining. In some industries there are joint labor/management trusts that administer pension, health and welfare programs.

Wage structure, progression, rates. This clause spells out wage rates by position, length of service, step rates, cost-of-living adjustments, and so on. Changes in wages are usually negotiated near the expiration time of the collective agreement, although some collective agreements have wage-reopener clauses during the term of the agreement.

Wash-up time. In situations where employees need to change clothes, wash up before or after working or use special uniforms, there is usually a wash-up period stipulated in the agreement that states the amount of time allocated and whether or not employees will be paid during this period (they usually are).

Duration of collective agreement. This spells out the exact dates covered in the contract, usually one to two years. By law (National Labor Relations Act) agreements covered by that Act can run no longer than five years.

Management rights. Management rights are also spelled

Figure 8 – 3. Typical Collective Agreement Rules

Virtually all collective agreements cover the following subjects:

- Parties to the Agreement
- A statement of purpose
- Recognition clause
- Coverage or Scope
- Union Shop
- Hiring Hall
- Check-off
- Discrimination
- Hours of work
- Overtime
- Shift premium
- Holidays
- Call-in and reporting time
- Vacation
- Leave of absence
- Seniority
- Method of wage payment
- Union representation, stewards and visitation
- Grievance procedure
- Arbitration
- Discharges
- Bulletin boards
- Jury duty
- New jobs
- Strike and lockout
- Waiver and alteration of agreement
- Safety and health
- New employees
- Death in the family
- Health and welfare
- Wage structure, progression, rates
- Wash-up time
- Duration of collective agreement
- Management rights

out in the collective agreement, and it is typically agreed that these items are not considered working conditions as described by the law for the purpose of collective bargaining. Usually, but not in all collective agreements, management rights specifically cover the following subjects:

- Work to be performed
- How work is to be performed
- Tools, equipment and machines to be used (except if a safety hazard is involved)
- Money to be spent in performing the work
- Company organization structure
- Selection of supervisory personnel
- Need for increase or decrease of employees performing the work
- Standard selection of employees

All of these management rights, of course, are to be taken in the context of the previously stated rules in the collective agreement (Figure 8 – 3).

CONTENTS OF A POLICY/PROCEDURE MANUAL

Company-written policy/procedure manuals and employee handbooks are defined here as those statements of terms of work, benefits and conditions of employment produced by the company for its employees. They will be called *policy manuals*. These exist in union and union-free workplaces. In union workplaces, such policy manuals are usually somewhat abbreviated, since many of the collective agreement rules cover the subjects normally included in a policy manual.

The contents of a company policy manual as used in a non-union environment are not much different from those of a labor/management collective agreement in terms of topics covered and even many of the procedures involved. At one time, these policy manuals were not considered as binding as a collective agreement. However, more and more state and federal courts are interpreting policy manuals and employee handbooks as binding contractual agreements. The major legislative changes are occurring at the state rather than the federal level. The concept of employment-at-will is rapidly eroding. The idea used to be that the employer had the unilateral right to hire and fire for any reason not specifically in violation of the law. Things just aren't that way anymore. The courts and state legislatures are rapidly moving toward what is often called the democratic workplace. The idea of due process in the workplace is replacing the idea of employment-at-will. Company-written manuals are now treated as contracts. The absence of a manual can be even worse, since management and employees then must rely on implied or disputed understandings. That's why the Supervisor's interpretation or administration of the manual is of critical importance.

Topics Included in the Manual

Most manuals address working conditions, such as vacation, wages, benefits, hours of work, leaves of absence, holidays, etc. Many policy manuals also include some type of grievance procedure although it may not be as specific or binding as those found in a union agreement. In fact, the five-step grievance process now being adopted and used by most nonunion companies is largely reflective of the form of grievance process found in most collective agreements. The primary difference in this and similar provisions is that in a collective agreement the employee has the right to be represented by a union official.

Policy manuals may well include programs that are union-like, such as employee/management committees to discuss issues and problems, complaint systems, counselling with assured anonymity, and so on. Some organizations set up these union-like structures to get employees involved and give them a sense of due process even though there is no formal union representation.

Topics Not Usually Included in the Manual

Policy manuals will not include items particular to a collective agreement, such as a recognition clause, union shop, check-off, discrimination as it relates to the union, union representation, stewards, strike, lockout, scope, waiver or alteration of agreement and duration of the contract.

Figure 8 – 4 is a comparison checklist of a union versus a nonunion environment in terms of the topics covered by collective agreements and policy manuals. Again, the differences come in the specifics: how a rule is written versus how a policy is written. How someone is represented versus the position that she/he does not need representation. How due process is applied. Some see these as very significant differences while others do not.

LEGAL CONSIDERATIONS

In a union working environment with a collective agreement, you as a Supervisor must realize that recognizing the union and following the agreement is more than the right thing to do, it's the law! The National Labor Relations Act, Section 1, sets forth the policy of the United States with regard to the right of employees to union representation and collective bargaining. It provides in part:

Experience has proved that protection by law of the right of employees to organize and bargain collectively safeguards commerce from injury, impairment, or interruption, and promotes the flow of commerce by removing certain recognized sources of industrial strife and unrest, by encouraging practices fundamental to the friendly adjustment of industrial disputes arising out of differences as to wages, hours, or other working conditions, and by restoring equality of bargaining power between employers and employees.

Several of the legal considerations regarding union agree-

Figure 8 – 4. Comparing Topics Typically Covered in Collective Agreements (CA) and Policy Manuals (PM)

Topic Covered (Y = covered; N = not covered; S = Sometimes)

CA	PM	Subject
Y	Y	Parties to the Agreement
Y	Y	A statement of purpose
Y	N	Recognition clause
Y	S	Coverage or Scope
Y	N	Union Shop
Y	N	Hiring Hall
Y	N	Check-off
Y	Y	Discrimination
Y	Y	Hours of work
Y	Y	Overtime
S	S	Shift premium
Y	Y	Holidays
Y	Y	Call-in and reporting time
Y	Y	Vacation
Y	Y	Leave of absence
Y	S	Seniority
Y	Y	Method of wage payment
Y	N	Union representation, stewards and visitation
Y	S	Grievance procedure
Y	S	Arbitration
Y	Y	Discharges
Y	Y	Bulletin boards
Y	Y	Jury duty
Y	S	New jobs
Y	N	Strike and lockout
Y	S	Waiver and alteration of agreement
Y	Y	Safety and health
Y	Y	New employees
Y	Y	Death in the family
Y	Y	Health and welfare
Y	Y	Wage structure, progression, rates
Y	Y	Wash-up time
Y	N	Duration of collective agreement
Y	Y	Management rights

ments were discussed in Chapter 7, such as not interfering with a union representation election and not discriminating against employees due to their union membership/activity. In addition, breaching the collective agreement, like breaching most contracts, can result in costly and time-consuming arbitration or even more severe disputes. Supervisors who generate a lot of grievances make few friends on the higher levels of management. It's simply a bad supervisory practice.

The term *management*, as used in the collective agreement, usually means in practical terms, the Supervisor. The Supervisor is the one closest to the employees and is responsible for the day-to-day administration of the collective agreement. Thus the Supervisor must help protect the employer from any legal liabilities by making sure that the agreement is administered fairly and effectively.

ADMINISTERING COLLECTIVE AGREEMENTS

Administering collective agreements has many implications for the Supervisor. Supervisors who have not worked in a union environment may fear the collective agreement, seeing it as a threat to the effective handling of their supervisory duties. On the other hand, many Supervisors who have spent their lives administering a collective agreement can't imagine doing their job any other way.

Why Employees Join Unions

Understanding why employees join unions may help put things in perspective. Some of the many reasons employees join a union include:

- Security of association with others in the same condition, i.e., safety in numbers
- Desire to increase or secure their share in the economic system
- Sense of independence and control over their own affairs
- Means to understand and deal with the forces and factors at work in one's world
- Socio-cultural heritage that assumes the need and right to belong to a union. It's the thing to do.

Those explanations come down to: strength in numbers, more money and benefits, dignity and freedom from intimidation, leverage and habit. Most people, sometime in their working life, have similar concerns in the relationship with their employer. Therefore, what is it that makes the difference? Why do some employees choose to go union while others do not?

The Key

The Supervisor is key to the implementation of management/labor relations. It is the Supervisor's responsibility to assure that the employer's rights are preserved in practice. The Supervisor must also assure that the conditions of employment agreed upon are carried out fairly and honestly. In other words, the Supervisor is also trying to assure that labor's rights are preserved in practice!

In the eyes of most employees, the Supervisor is management. They will often look to the Supervisor for interpretation of the collective agreement rules. Therefore, the Supervisor must first know what is in the collective agreement and any other employer policies and procedures. The agreement cannot be ignored. It must be followed as it was intended to be followed.

To effectively administer the collective agreement, you will have to get to know the union representative responsible for the people working for you. Sometimes this is a steward, a business agent, or whatever title the union uses. Get to know the union representative or steward, and strive for a good working relationship. Keep in mind that this individual is selected by the employees and works on behalf of the union. Sometimes, the person may also be doing double duty, working full time for the company and part-time for the union. Good communication with the union is invaluable, and most of the union contact you will have will be with the local employee representative.

In addition to the specifics of the union agreement that the Supervisor must administer, the best way to avoid breaching the contract or the filing of grievances is to apply sound supervisory skills. These skills have been discussed throughout this book, but here are a few highlights:

- Strive at all times to bring understanding rather than confusion
- Be consistent in all that you do
- Communicate your expectations and act as a leader—what is expected of a Supervisor
- Realize that what you get is usually what is given
- Live up to your work
- Treat employees as individuals and respect their differences
- Learn to communicate with your employees: ask open-ended questions, listen actively, talk in language the other person understands
- Show enthusiasm about your employees
- Give feedback and show appreciation
- Personalize good things and depersonalize bad things
- Help others be successful
- Be fair, honest and impartial in the application of all rules and policies

A lack of understanding between Supervisor and employees is the greatest single basis of grievances. Does this mean that the number of grievances being filed in your area reflects on you? Yes, it does. At the same time, you do have to apply the agreement and get the work done. The best way to do that is to get to know your employees and keep them informed. Also keep in mind that the day-to-day administration of the collective agreement influences what will be brought to the bargaining table at contract negotiation time.

For example, if in choosing between two equally qualified

employees for a promotion, you choose the one with less seniority and who also happens to be seen as your favorite, chances are seniority systems will be brought to the bargaining table the next time the contract is up for negotiation. And between now and then, you'll have a lot of grievances over the issue.

Fair treatment of and good communication with your employees under a collective agreement can also have safety implications. Let's look at one such instance.

CASE STUDY: A UNION ENVIRONMENT

A maintenance man at Johanson's, a local department store, thought he had been treated unfairly by his Supervisor. Not only was he getting ready to file a grievance, but this situation also had safety ramifications.

The Situation: When revolving doors are folded back into the center of the door opening to give clear passage, metal screw plugs must be removed from the floor sockets so that the locating rods that hold the doors open can drop into them. If these plugs are not replaced when the doors are again unfolded (for use as a revolving door), the open holes are a hazard for women with high heels. If heels get caught in the holes while doors are revolving, someone can be seriously injured. Frank, the maintenance man, had not replaced the plugs, and Owen, his Supervisor, had received a number of complaints. The Supervisor had asked Frank to come to his office.

The Conversation:

Owen: *Come in and sit down, Frank. I want to talk with you.*

Frank: *OK, boss. Is something wrong?*

Owen: *Not wrong, Frank. It's something that's up to us to watch out for to protect our customers.*

Frank: *What?*

Owen: *It's those floor plugs in the revolving doors. They weren't put back yesterday and a couple of women caught their heels in them.*

Frank: *Oh, that. You know, Owen, some days I feel I just can't suit you. Maybe you ought to get yourself another man. I'm not getting anywhere around here, anyhow.*

Owen: *Now, Frank. I like your work generally. Is something in particular bothering you on the job?*

Frank: *Yes, and I'm fed up.*

Owen: *Tell me what's wrong.*

Frank: *I've been at Johanson's going on two years as mechanic, and all I do is odd jobs. You gave Pete, who has six months less time, a machine maintenance job, but not me. And he filed a grievance last month about his overtime pay. Looks like you're playing favorites, Owen—and here I am working my fingers to the bone. Maybe I should just file a grievance, too.*

Owen: *First of all, Frank, if you've got something to file a grievance over, it's your right to do it under the contract. I don't have to tell you that. But I would rather work it out* with *you, if possible. Is your complaint that Pete got the promotion?*

Frank: *Yeah, of course.*

Owen: *Didn't you know? Pete had three years of machine maintenance work before he came here. I had to have a man quick. There wasn't enough time to train someone.*

Frank: *Oh, I see. It looks just like I said: I'm getting nowhere here. I didn't have the background, and so I'm just stuck.*

Owen: *Wait a minute, Frank. I've already made arrangements for you to get five hours' training a week so when the next opening comes up, you'll be eligible. I'm sorry I didn't have a chance to tell you yet, but it hasn't been exactly slow around here today.*

Frank: *You have? Well, thanks Owen.*

Owen: *That's OK, Frank. Now, about those floor plugs. I know you and the other men don't skip them deliberately. They are small and get lost or misplaced.*

Frank: *But Owen, no one's been hurt yet.*

Owen: *No, not here. But at Wellington's store, a customer caught her heel in an unplugged hole, the revolving door was moving fast and hit her before she got loose. She broke her ankle.*

Frank: *I hadn't heard. Fill me in. What happened, Owen?*

Owen: *Well, the store heard plenty. A big lawsuit is being filed. I wish I knew the answer, Frank, for how we could avoid a similar situation.*

Frank: *I've been thinking, Owen. Suppose we take a piece of wide adhesive tape and stick the two plugs to the door. They're pretty small.*

Owen: *I think you have a good idea, Frank. It may not look too good on those bronze doors, but it will work.*

Frank: *Yes, it wouldn't look too hot and some kid might peel it off, too. Here, I've got it. A small matching metal box high up on the door with a spring cover, too high for kids to reach.*

Owen: *Now you've got it. That's it. Good work. I wish they were on now. Here, make me a sketch of what you have in mind.*

Frank: *OK. (Frank draws a sketch.) See, I put on the dimensions.*

Owen: *Which side is mounted up?*

Frank: *With the spring down, see? I'm sure the shop can make up the boxes, and I'll get them on in a day.*

Owen: *OK, Frank. You see the shop foreman. I'll tell him to get you what you want and give him your sketch.*

Frank: *Owen, we can tape the plugs on until I get the boxes installed. The doors are folded open today, so I'll do it now.*

Owen: *Real good idea. Thanks, Frank!*

If Owen had not gone the extra distance and kept trying to resolve the personal work concerns Frank had, the situation might have deteriorated. A grievance might have been filed. Frank may not have offered a solution to the problem with the door plugs, and an accident may have occurred. In addition, communication and understanding would have decreased. Because Owen took the time to talk with Frank, a grievance filing was avoided, open communication was fostered and a safety hazard was eliminated.

Administration of the union agreement also has implications for the productivity of the organization. Adversarial relations may be inevitable at contract negotiation time. However, they need not be carried into the day-to-day workplace. Most workplace issues can be resolved with effective communication and understanding. They do not need browbeating from either management or labor.

Freeman and Medoff, both management people, in *What Do Unions Do?,* carried out an exhaustive study and comparison between union and nonunion work environments. The surprising results of their study were the general lack of differences between the two types of environments. One of their conclusions makes this very clear:

> Unionism can be a plus to 'enterprise efficiency' if management uses the collective bargaining process to learn about and improve the operation of the workplace and the production process.... On the other hand, if management responds negatively to collective bargaining (or is prevented by unions from responding positively), unionism can significantly harm the performance of the firm.

The issue, then, is not so much whether the workplace is unionized or union free. The issue is how management and labor get along.

CONCLUSION: LOOKING BACKWARD AND LOOKING FORWARD

This chapter makes an important point that can be best understood when you have covered the full range of what a Supervisor does. Good supervision is good supervision. Effective employee/management relations have more to do with effective Supervisor/employee relations than any other factor. More than whether there is or is not a union, whether there is or is not a good relationship between a Supervisor and his/her employees will determine all the issues so critical to management. These include safety, productivity, morale, earnings, profitability and competitiveness. The relationship between the Supervisor and the employee are the foundation of all of these issues.

It is very clear that the Supervisor/employee relationship itself must be built on the foundation of some very fundamental concepts of the due process, democratic workplace: integrity, fairness, mutual respect and open communication. We have laws that espouse and attempt to enforce these concepts. What we need are more Supervisors, managers, employees and unions who will apply them. That is something to really look forward to.

Chapter 9

Interviewing, or Talking It Over

OVERVIEW

This chapter focuses on the skills necessary for a Supervisor to become a successful interviewer. We will discuss why it is important to become a skilled interviewer and what types of interviews are normally carried out by Supervisors. We will focus particularly on the employment or hiring interview, since the other types of interviews are discussed extensively in other chapters of the book. We will also look at accident investigation because it also requires interviewing skills. We will review the three different types of interview techniques and discuss the hiring process in general. We will then describe some practical right and wrong ways to do things in employee interviewing.

THE IMPORTANCE OF INTERVIEWING

Interviewing employees is one of the most important duties of a Supervisor. But, you say, *I don't do much interviewing. That's done by Personnel.* Wrong. When you sit down and talk with an employee about a job change, that's an interview even though you're just talking it over. When you talk things over in a performance appraisal discussion, that's an interview. When Personnel sends a potential new hire to you and you have a chat, that's an interview. The fact is, you do a fair amount of interviewing when you think about it this way (Figure 9 – 1).

At first glance, interviewing may sound like a pretty run-of-the-mill supervisory activity. Such an impression is far from reality. A Supervisor with good interviewing skills actually impacts the workplace in several ways. As the saying goes,

The better the interview, the better the person hired; the better the person hired, the better the productivity and safety; the better the productivity and safety, the better the organization.

Supervisors conduct many different types of interviews (Figure 9 – 2). Interviewing an applicant for a job is one type of interview. The variety of interviews carried out by Supervisors that require similar skills include:

- Exit interviews (when employees leave the company)
- Reference checking
- Performance appraisals
- Discipline interviews (including safety violation reviews)

Figure 9 – 1. Interviewing: Why Is This So Important?

- Job promotions
- Accident investigations
- Employee safety counseling

In this chapter, we will be dealing primarily with the skills and conditions of employment or hiring interviews, exit interviews, accident investigations and employee safety counseling. We have also included a case study of employee safety counseling. Chapter 10 deals with performance appraisal interviews and Chapter 11 with discipline interviews.

THE HIRING PROCESS

Hiring a new employee is a multistep process that involves the applicant(s), the organization, Human Resources or Personnel and you, the Supervisor. The employment interview (conducted by Human Resources and/or the Supervisor) is a large piece of the process, but keep in mind the other steps that go along with it. Each organization has its own policies and procedures, but almost

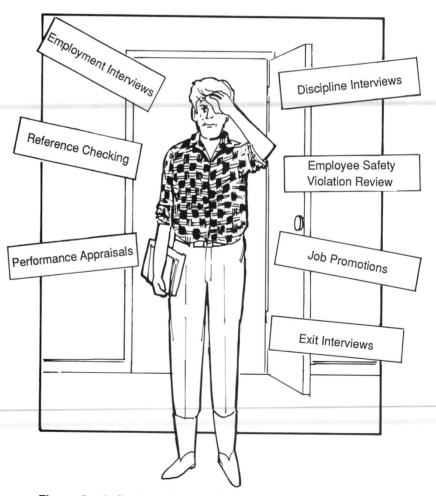

Figure 9 – 2. So Many Types of Interviews

all include an interview between the Supervisor and the proposed new hire.

Many Supervisors see their part in this process as very small. That just isn't true. When a potential new hire is sent to you for an interview, your judgment as to suitability for the work involved is the single most important part of the whole hiring process. If you don't exercise it carefully, everyone's in trouble.

Getting Ready

There are a lot of things you should know, and a lot of information you should review before you ever interview anyone for a job. Here are some guidelines (Figure 9 – 3).

1. Become familiar with your organization's hiring policies. Your Human Resources Department may have very specific guidelines to follow for filling a vacancy. A systematic and consistent approach to filling jobs is the only really reliable approach.
2. Identify the position for which you are hiring and the qualifications necessary to succeed in the position. This means you should do the following:
   - Locate the job description, and review it for accuracy and up-to-date information.
   - Determine the actual skills, experience and education necessary for the job, and rank their importance in relation to each other. For example, if experience is more important than years of education, say so.
   - Avoid adding qualifications that are not absolutely necessary for the position.
   - Identify and rank personal characteristics that are necessary for the job, such as initiative, attention to detail, and/or ability to get along with others.
   - Identify elements of the work environment of the job, such as many deadlines, tight time frames, safety considerations, split shift work or overtime.
   - Take advantage of any training offered by your company regarding your organization's hiring process.

Finding Applicants

Finding good job applicants is a critical part of the process. Some Human Resource Departments retain responsibility for soliciting applicants, and you may not have to conduct this step in the hiring process. If you are not involved, it is still in your best interests to advise Personnel of organizations you know that would be good places for employment opportunity notices. If you are involved in the process of seeking applicants (and this is the trend), you need to tap into your known resources for a good applicant flow. How do you do that?

First, use your own resources. Post notices within the organization. Network within your circle of employees or with professional associates. Place advertisements in local newspapers, neighborhood organizations, community action organizations, schools, colleges or universities, affirmative action networks and trade associations.

Figure 9 – 3. The Hiring Process: Getting Ready

- KNOW your organization's hiring policies
- KNOW the position for which you are hiring
- KNOW the qualifications necessary to succeed in the position

Second, check local labor union resources. In some instances, hiring from the union hiring hall may be required under the terms of the labor agreement between the union and the company. This is particularly true of the skilled trades and the construction industry. In other cases, the relationship between management and the union is positive, and the union may frequently refer potential applicants. In any event, be sure to check your organization's policies on this.

Third, seek out government resources. State employment services maintain current rosters of available workers. Equal Employment Opportunity networks often provide referral services. Most companies require you to post your notices with these agencies. Some companies are obligated under affirmative action orders to use these programs extensively.

Fourth, seek employee referrals. Some organizations have incentives for this type of referral while others actually discourage referrals. Again, know the policy of your organization (Figure 9 – 4).

Screening Applicants

Like recruiting, screening is a process that some companies retain in Human Resources. However, the trend is to involve Supervisors in the process. Before you can do any screening, you need to be familiar and comfortable with company policy and government regulations regarding minorities, affirmative action, handicapped and protected classes generally. The legal part of this was covered in Chapter 7. Company policy is probably printed in a Personnel or Employee Manual. Having reviewed that, you move on to screening. As a result of the recruitment process, you will in all likelihood have more applicants than you know what to do with. Since interviewing each applicant would completely divert you from your main Supervisory work, it's time for some sorting and selecting.

The first thing you need to do is screen, which means thinning things out (Figure 9 – 5). Sort, on a preliminary basis, to see which people you will interview. How do you do this? Take a good hard look at all of the applications by reviewing each application and/or resume. Ask yourself these questions to help you select which applicants you may want to interview:

- Is the application complete? Note any employment gaps or omissions.
- Is the overall appearance of the application legible and neat?
- Does the applicant possess most of the credentials necessary for the job?
- Note any special training the applicant has had.
- Review for any overlaps that don't make sense. Attending school and working at the same time is quite possible, but notice if the school and work are in the same city, for instance. Note volunteer work that may explain gaps, and that also may have transferable skills. For example, volunteer typing and office work for a church society may well

Figure 9 – 4. The Hiring Process: Finding Applicants

If this is part of your supervisory duties, find applicants by:

- Posting notices in the organization
- Checking with your employees and associates
- Advertising locally
- Checking the local union hiring hall
- Checking the union for referrals
- Using state employment services
- Using employee referrals (if company policy allows)

Figure 9 – 5. The Hiring Process: Screening Applicants

Too many interviews defeat the process. You must screen your applicants and select those who appear to qualify. Use this checklist to thin out the list:

_____ Is the application complete?

_____ Is the application legible?

_____ Does the applicant possess most of the skills and credentials necessary for the job?

_____ Has the applicant had any special training?

_____ Are there overlaps which don't make sense?

_____ Are there inconsistencies?

_____ Did the applicant change jobs a lot? Check for reasons.

_____ Note the reasons given for leaving previous employment.

Based on all of the above, determine who looks the most qualified and create a pool of those you would like to interview.

show the ability to type, compose letters, handle mailings, etc.

- Look for any inconsistencies. For example, do the credentials suggest that the applicant is overqualified for the positions they have held? Is the salary history consistent with the positions held?
- Check for the frequency with which the applicant has changed jobs. There may be legitimate reasons for frequent changes due to the nature of the industry involved, or it may reflect that the person has difficulty holding a job.
- Note the reasons the person has listed for leaving previous employment. Generally, keep notes on a separate sheet of paper, possibly attached to the application with a staple.
- Based on all the above, determine who looks like the most qualified and create a pool of those you would like to interview.

Testing

Pre-employment testing is conducted by Human Resources, when it is used at all. Many companies rely entirely on the interview assessment process instead of testing. Testing has always been subject to close scrutiny by Equal Employment agencies, since it frequently results in a disproportionately adverse impact on hiring from protected classes. But some companies do test. Some companies share testing data outside the Human Resources Department. Most do not. Increasingly, testing focuses on two qualities: intelligence and trustworthiness.

Specific skill-related tests are seldom given, with companies relying instead on certifications and diplomas. This is because in the past there was a tendency to make skill-related tests particularly difficult on issues related to lifting and physical prowess in general. There has also been a tendency to only apply these tests to women or smaller males because *I thought they looked like they couldn't handle it.* For example, it is discriminatory to insist that the applicant must be able to lift and carry a 100-pound steel rail 30 yards when, in fact, that job is routinely carried out by two people. If you have such a test, it must be justified in terms of the actual needs of the job and must be applied to each and every applicant for that job, regardless of size or gender. Otherwise, such testing clearly violates Title VII of the Civil Rights Act. Even police and fire departments are eliminating many of the excessive physical test requirements they have had, with no apparent adverse impact on the quality of people being hired. If you use skill tests, do so only with the full approval of Human Resources.

The use of more exotic testing, such as polygraphs, has been severely restricted outside of government or security/financial jobs. Some companies may require applicants to take a medical examination after selection but prior to hiring. Such examinations often include tests for substance abuse.

Do not, under any circumstances, develop your own test, written or physical, for screening potential new hires. It's the job of Personnel to put tests in place.

Since all of this testing relates to pre-employment and occurs prior to the person coming under the direction of the Supervisor, we will not go into the matter in any further detail.

INTERVIEW METHODS

Before you interview an employee, you should have some idea of what you want to know and how you will find out. We will cover these matters in this section. But, you say, *I have always just talked with them!* Maybe you learned what you needed to know, and maybe not. But there are some methods that will make it a lot easier for you to find out what you need to know about a job applicant. So, the first step is to determine the type of interview technique you will be using. There are three types of interviews: structured, semistructured and unstructured (Figure 9 – 6).

Structured Interview

In the structured interview, the interviewer knows exactly what is to be found out and how it will be solicited. You come to the interview with a prepared list of questions that are asked, in full, of each and every candidate. This method is the most valid and reliable because you will be comparing answers to the same question: apples and apples, not apples and oranges. The structured interview technique may be an easier method if you do not interview frequently because with this method you know ahead of time how the course of the interview will go. It brings out the information you need. The structured interview is rigid in format, although the way people respond to the questions will vary widely. It also virtually eliminates biased questions from inexperienced interviewers and is, therefore, particularly popular when there have been problems with interviewers giving the wrong impression about company policy, affirmative action and so on.

Semistructured Interview

The semistructured interview technique calls for some prepared questions to be used for all applicants and other impromptu questions that are developed by the interviewer as the interview progresses. This method allows for more flexibility to pursue, for example, a particular expertise of the applicant. It may be less reliable for making a hiring decision if the impromptu questions are not asked of all candidates. The impromptu questions can also lead to diversions in the interview that cause it to drag on, sometimes to the extent that you do not get all of the information you need.

THE STRUCTURED INTERVIEW

You know exactly what is to be found out and how it will be solicited.

THE SEMISTRUCTURED INTERVIEW

You have some prepared questions and develop others based on applicant's responses.

THE UNSTRUCTURED INTERVIEW

You go with the flow.

Figure 9 – 6. The Hiring Process: Interview Methods

Unstructured Interview

Often referred to as the Fly-by-the-Seat-of-Your-Pants Method, this is probably what you have been doing if you haven't had some direction or training. No questions are prepared ahead of time. The interviewer asks questions based, at least, on a review of the application. This is the least reliable type of interview technique because the same questions are not consistently asked of each applicant, the interviews tend to wander and the interviewer tends to do all the talking.

As a suggestion, a good beginning would be to try structured interviews for the applicants for a particular job. After that experience, you might try a semistructured interview format for the next position. *We do not recommend the unstructured interview under any circumstances.*

PLANNING AND CONDUCTING INTERVIEWS

The detailed planning stage must occur before the interview takes place. Be clear about your objectives and company policy on hiring, firing, performance and discipline. Here are some major guidelines you should follow:

- DO prepare and plan for the interview. Review, once again, the key or critical qualifications you are looking for in an applicant.
- DO prepare questions ahead of time for use in the structured or semistructured technique. (DON'T use the unstructured technique.)
- DO prepare questions that are job-related and will bring out information regarding the key qualifications you have identified.
- DO develop questions that will elicit examples of past job performance.
- DO prepare questions around hypothetical situations for which the applicant must find a solution. *What would you do if there were a fire and...?*
- DO prepare open-ended questions and not just questions that can be answered with a yes or no response. For example, *When you worked for ABC, Inc., what did you do about...? How did you...? Why did you ...? Tell me about....*
- DO prepare questions that touch upon various areas of experience, such as education, military service, work experience, career goals/ambitions, assessment of one's own strengths and limitations and upon attitudes toward current job, safety regulations, equipment, company and industry.
- DO prepare probing questions. Probing questions are the only way you can get most people to open up and tell you about themselves. After all, people are tense when it comes to an employment interview, and who can blame them? But it is your job to find out about them and make an assessment as to their suitability for the job and compatibility with your workers. You will live with your mis-

takes. Probing questions are the best tool available to minimize those mistakes. Examples of probing questions are shown in Figure 9 – 7. Try answering them yourself in the space provided. Notice how you really have to struggle with some of these? That's precisely what the applicant will do, and why you will learn more from such questions.

Interview Tips

At last, we move to the actual interview. You and the employee or potential employee are about to enter the interview room. Sometimes (many times) there is no room, and often there isn't even a quiet corner. Then, you must make do as best you can. Stay away from others by going off to the side or going to the cafeteria. Do whatever is possible to create an environment in which your mutual attention will be on each other (Figure 9 – 8).

Here are some guidelines for conducting successful interviews:

- DO make the applicant as comfortable as possible in the interview. Develop rapport with the candidate. Be friendly and welcoming. Free yourself from any distractions by closing the office door, forwarding your phone calls and letting others know that you are not to be disturbed. Arrange comfortable seating. Concentrate on the applicant by having good eye contact.
- DO allow enough time to complete the interview at a relaxed pace. Rushing is very apparent to all and will cause the applicant to wonder why you even bothered to hold the interview. *(If you weren't serious, why did you have me waste my time coming here?)*
- DO keep in mind that you are representing the organization. Project a favorable, professional, friendly image and understand that in the applicant's eyes you are the company.
- DO have an introduction at the opening of the interview. This introduction should describe the interview process and, briefly, the position the individual is applying for. It should also describe the organization and your particular department within the organization.
- DO ask the questions that you have prepared ahead of time. Ask the easier questions first and progress to the more difficult questions. This helps to put the applicant more at ease. Don't divert yourself or allow the applicant to divert you. Stick with the program.
- DO allow time for the applicant to respond. Remember, the purpose of the interview is to listen to the applicant.
- DO allow for pauses or silences; the applicant may offer additional information. Pauses are sometimes difficult for some interviewers, and they tend to fill in the void with their own talk. That is a mistake. A well-timed silence or pause will often cause the applicant to offer you more information of value.
- DO listen (actively) more than you talk.
- DO probe for further information if the applicant doesn't

Figure 9 – 8. Interview Guidelines: DO

Remember the KISS principle? Applying it to interviews:

DO:

Make the applicant as comfortable as possible.

Allow enough time to complete the interview.

Remember whom you're representing.

Prepare an introduction.

Ask prepared questions.

Allow time for the applicant to respond.

Use pauses or silences.

Listen actively.

Probe for further information.

Restate responses.

Make your questions clear.

Take cursory notes.

Be sensitive to cultural differences.

Have a conclusion to your interview.

Evaluate the information and make a hiring decision.

Make reference checks before making an employment offer.

Figure 9 – 7. Ask Probing Questions

Try applying this to yourself by answering a few of these questions briefly in the space provided. See what happens!

• On your last job, what was it you most wanted to accomplish, but didn't?

_____
_____
_____

• What was the best job you ever had? The worst? The best boss? Worst boss? Why?

_____
_____
_____

• What do you pride yourself on?

_____
_____
_____

• Why are you interested in changing careers?

_____
_____
_____

• Why do you want to leave your present position?

_____
_____
_____

• What was the problem between you and your last Supervisor?

_____
_____
_____

• What kind of person do you most (least) like to work with?

_____
_____
_____

• What areas of your work need improvement?

_____
_____
_____

• If I were to call your immediate past employer or immediate Supervisor, what would he/she tell us about you?

_____
_____
_____

• Do you think you should be able to criticize management?

_____
_____
_____

• How do you react to rumors on the job?

_____
_____
_____

Figure 9 –7. Continued

• What experiences do you think you have carried from your education and applied in your work experience?

_____

_____

• What would you say is the most important thing you learned from your education/military experience/last job?

_____

_____

• What changes have you ever made in your approach to others in order to become better accepted in your work setting?

_____

_____

• What would you say are the most important things you are looking for in an employer?

_____

_____

• What are some of the problems you encounter in your job?

_____

_____

• What frustrates you the most? What do you do about it?

_____

_____

• How do you feel your current boss rates your work? What are some of the things he/she indicated you could improve upon?

_____

_____

• What do you think are the most important characteristics and abilities a person must possess to be successful in the position you are applying for? How do you rate yourself in these areas?

_____

_____

initially answer the question as completely as you'd like. Probe further if the applicant seems hesitant or nervous about a particular situation. If you don't probe at these points, you may overlook some very critical information.

• DO restate responses from time to time to make sure you are understanding the applicant's answers to the questions. For example, _If I understand correctly, Jane, you left your previous job because the tension was too great, and you prefer a more relaxed atmosphere?_ Or, _I'm still not certain, Kim, about where you were working between December, 1987, and March, 1988. Would you go over that again, please?_

• DO make your questions clear and restate them if the applicant doesn't understand the questions.

• DO take cursory notes. Unless your memory is phenomenal, you'll need them.

Figure 9-9. Interview Guidelines: DON'T

This list is best reviewed in advance of an interview. If you wait until after the interview, you'll just get that sinking feeling.

DON'T:

Go into an interview unprepared.

Ask illegal, irrelevant or dumb questions about sex, race, color, religion, age or disability.

Ask about skills or abilities not involved with the job.

Exaggerate the job responsibilities, career opportunities or salary information.

Talk more than you listen.

Interrupt the candidate.

Jump to conclusions.

Like someone because he/she was sent by someone you like.

Dislike someone because he/she was sent by someone you don't like.

Make a decision based on initial body language.

Make a positive or negative decision based on your own values, standards and beliefs rather than job issues.

Concentrate more on reading the application or resume during the interview.

Telegraph answers.

- DO keep in mind the cultural differences explained in Chapter 6 when communicating during an interview. Many interviewers turn down well-qualified people because they have a little difficulty understanding them or think language is critical when it really isn't for the job involved. Remember how much steel was built and how many rails were laid by people who couldn't speak a word of English.
- DO have a conclusion to your interview. Ask the applicant if he/she has any questions. Explain the next step(s) in the hiring process. Be sure to thank the applicant for his/her time.
- DO evaluate the information you have gathered and make a hiring decision. Exchange and compare notes with anyone else who interviewed the candidates. Have additional interviews or interviewers if additional information is needed. Focus decisionmaking on the critical requirements for the job that you identified initially. Don't start redefining jobs to fit particular individuals.
- DO make reference checks before extending an employment offer. People do occasionally lie when they apply for a job. Remember that employer reference checks are usually very guarded, with current practice generally limiting responses to dates of employment and title.

Interview Pitfalls

This is a very extensive list of all the things you shouldn't do, unless you don't mind messing up an interview or being named in a lawsuit (Figure 9 – 9).

- DON'T hold an interview if you are unprepared, don't have enough time or are too distracted to give the applicant your undivided attention.
- DON'T ask illegal or irrelevant questions. Hiring decisions cannot be based on one's sex, race, color, religion, age or handicap. Therefore, don't ask questions like:
  Where were you born?
  How old are you? (You may ask age after the person is offered the job.)
  What is your overall medical condition?
  Are you married?
  Do you have or plan to have children?
  How will you manage your children?
  How can you travel when you have children?
  What church do you belong to?
  Have you ever been arrested? (You may ask about convictions, but this information should only be considered if relevant to the job at hand.)
- DON'T ask questions that have no relevance to the position requirements. Examples: Don't ask about the applicant's typing skills if typing is not necessary in the position. Don't ask if the applicant can carry a 150-pound rail alone if it is normally handled by two people.
- DON'T exaggerate the job responsibilities, career oppor-

tunities or salary information. If the applicant takes the job, and it's all a fairy tale, you'll soon have a very disgruntled employee.

- DON'T talk more than you listen and DON'T interrupt the candidate while he/she is talking. Does this sound familiar? It is familiar because we've repeated it a lot. It is the single most common mistake made by interviewers.
- DON'T jump to conclusions or give in to your prejudices or biases. DON'T make a decision prematurely. First impressions play a role but should not exclude additional factors.
- DON'T like someone just because he/she was referred by someone you like, and DON'T dislike someone just because she/he was referred by someone you don't like.
- DON'T make a decision based on initial body language. Some people are initially very uncomfortable in an interview. Give them more time.
- DON'T make a decision based on your own values, standards and beliefs, either positively or negatively. For instance, someone who graduated from the same high school you did should not be a shoo-in for the job. Remember how many fellow graduates you really disliked?
- DON'T concentrate more on reading the application and resume or taking notes than on listening to the candidate. In most cases, the applicant expects that you have already read the application. If you haven't, you should have.
- DON'T telegraph, that is, don't give information or ask questions in such a way that the applicant can readily determine the responses you want to hear. For example, a question like, *You can handle stressful situations can't you?* will produce a *Yes, of course!* People will try very hard to give you what they think you want. You're not hiring you.

ADDITIONAL TYPES OF INTERVIEWS

As noted earlier, there are many types of interviews that a Supervisor conducts as a part of the job. We have concentrated in the previous section on the employment interview and will deal with such things as performance appraisal interviews in Chapter 10, discipline interviews in Chapter 11, interviews in the planning process in Chapter 12 and safety counseling interviews in Chapter 13. There are two other forms of interviews, however, that complement the employment or hiring interview: checking references and exit interviews. In addition to these, we will look into accident investigations as an example of the full use of Supervisor interviewing skills.

Checking References

As with hiring interviews, the question of who conducts reference checks is usually spelled out in company policy. Sometimes that function is carried out by Human Resources and other times by the Supervisor. Increasingly, this work is being added to Supervisors' duties. However, check your organization's policy first before you start doing this! If your organization does not have Supervisors do this work, you will still find the following information useful to better understand how people are hired and why some are not hired.

Checking references may not seem to be a form of interview, particularly since you will not be talking to an employee or a potential employee. However, you will be talking to people in the process of checking, and the best approach to that process is to use an interview method.

There are two very simple and very important reasons for checking references (Figure 9 – 10).

1. To verify that the applicant has truthfully presented his/her work experience, credentials and character.
2. To protect the employees and the company.

Unfortunately, we hear more and more about people falsifying their credentials and getting away with it for years because no one has bothered to check references. A recent survey of placement professionals suggested that 30% of all resumes/applications contain false or exaggerated information. If that is true, companies may be hiring a lot of people who aren't qualified for the job involved, or who may even be hiding something very detrimental to their future performance. Despite a million explanations, excuses or exceptions, one's past performance is usually an excellent indicator of one's future performance. Past Supervisors or colleagues can offer a lot of valuable information on the subject. You will need that information.

There are a growing number of court cases in which employers are being held liable for employees who may commit grievous acts while on the job or performing work-related responsibilities. When it involves a new hire, this is often called negligent hiring. For example, an employer was recently found liable for an employee who had raped and murdered a visitor to the employer's facility. The employee who committed the crime had been convicted and imprisoned for abduction and murder prior to being hired by the employer. The employee was out on parole. Nobody checked the references. A jury awarded $5,000,000 to the murder victim's family, payable by the employer. If you were the employer, what would you do with the Supervisor who didn't check the employee's references?

How to Check References

Assuming your organization's policy is that the Supervisor checks references, there are some basic steps and procedures that should be followed.

Figure 9 – 10. Reasons for Checking References

FIRST PURPOSE: to avoid hiring liars

SECOND PURPOSE: to protect the organization from potential liability.

- Secure from the applicant a signed release of information that allows you to check references, past employers, education, etc.
- Remind the applicant that a firm offer of employment cannot be given until a reference check is successfully completed. Putting it straight: if the applicant won't let

you do a reference check, you won't be offering him/her a job.

- Compile a list of the individuals whom you will be interviewing as part of the reference check. Current and previous Supervisors are probably the best source. As you know, a Supervisor knows the individual's performance and capabilities. The Supervisor may also be a good resource to determine names of other individuals familiar with the applicant. Fellow employees may also be good resources for a reference check. The Human Resources Department of the organizations the applicant has worked for will be able to verify dates of employment, salary and title information. Beyond that, however, they probably won't be able to give you specific information on the applicant's performance. Personal references listed on the application (most often teachers, relatives and friends) should be contacted but probably won't be as valuable as the other references. After all, applicants rarely list as references individuals who would make negative comments.

- Prepare questions that you will want to ask references. Keep in mind the same steps that have been used to prepare hiring interview questions. Many of the questions, in fact, are identical, although from a different point-of-view. For example, *What would you say are Tanya's major strengths?* instead of *What are your major strengths?* You should take particular care to target questions that address technical skills, work attitudes, interpersonal skills, communication skills, past accomplishments, future potential and character.

- Begin with positive questions so the reference person you are interviewing is not put on the defensive. Most people want to spread good news about a fellow or former employee. Give them that opportunity before you move on to the hard stuff.

- Keep your questions direct and concise. Figure 9 – 11 shows some fairly standard examples of what you need to ask, how you ask it and what you do with responses. It includes and expands on the questions asked in the case study that follows.

- If you don't get a clear answer, rephrase the question and try again. Sometimes you may have to push a bit, and sometimes you simply will not get an answer. Most often, the latter means potential trouble. However, many companies now have policies that forbid Supervisors to give references on past or current employees. You may want to ask if that is the case.

- References may be checked by mail, over the phone or in person. Checking references by mail is the least effective method because you may not get a response, and if a response does come, it may not be as candid as a personal reference check. Doing reference checks in person is the most effective but may not be feasible because of the time involvement. It is most likely that phone references will be the most common technique you will use.

- Always identify yourself, the organization you represent and the individual you are seeking information about. Let the reference know that you have the applicant's written consent to do the reference check.
- Be professional and courteous at all times as you ask the questions you have prepared.
- Keep in mind the nature of the source. For instance, the most recent employer may still be resentful of the individual for leaving and thus may not be very complimentary. (That is why it is so important to ask specific questions about performance, so you get more than just opinions.)
- Document the reference checks. Note the names of the references checked, the date of the interview and the questions and responses.
- If the applicant's educational background is relevant, you will want to send for a transcript. Many educational institutions will also verify education over the phone. The Principal's Office in a high school and the Registrar's Office in a college should be contacted for these records. You should be prepared to provide the individual's name during enrollment at the school, (the fact that names change is a frequently overlooked matter in reference checking) dates of enrollment and dates of graduation. Date of birth and the applicant's social security number may also be necessary to verify this information.

Figure 9 – 11. Checking Out Jerry: Examples of Reference Check Questions

These are a few of the questions that were asked about Jerry, and a lot more that would have been asked if the interview had gone differently. Use these as a checklist when you're doing reference checks.

- What were Jerry's dates of employment, title, salary?
- What were Jerry's responsibilities?
- How much supervision does Jerry require?
- Tell me three to five strengths or limitations Jerry has?
- How quickly does Jerry learn?
- Did Jerry have any work-related conflicts?
- How does Jerry compare with others who have the same responsibilities?
- What is Jerry's overall attitude toward his job, boss, fellow employees?
- How does Jerry present himself within the company?
- How does he handle stress?
- Was he involved in any work-related accidents?
- Did Jerry follow work rules, particularly safety rules?
- Did he like to work as a team or individually?
- What additional training does Jerry need?
- Why did he leave the organization?
- Is Jerry eligible for rehiring?
- Who else is familiar with Jerry's work skills?

CASE STUDY: CHECKING OUT JERRY

Gerald Robinson is at the top of your applicant list. You have listed him as your first choice for the Lab Assistant I position, requiring a high school diploma with a concentration in chemistry and science and previous metallurgical research experience. You had a good session with him at the interview and have done some work finding out who his former Supervisors have been. You are now talking to Carolyn Squire (CS), his former Supervisor at ACME Metals.

YOU: *Hi. I'm Bill Calder, Supervisor from Division One of Metallurgical Research Services, and I'm considering offering a position to Gerald Robinson who used to work for you. I told Gerald I was going to talk with you, and he has given me his written permission to do a reference check. I wonder if you would take a few minutes to talk with me about Gerald?*

CS: *Sure, I'll be happy to. Jerry was a great employee who worked for me a long time. I really like Jerry and wish him well with you. You've got a great employee!*

YOU: *Thank you, thank you. But I have a few specifics I need to get. Can you tell me what were the dates of employment when Jerry worked for ACME?*

CS: *Well, not exactly, I don't have those records.*

YOU: *Approximately, then?*

CS: *Oh, from about February, 1986, to last month.*

YOU: *Thanks. And what was his title and his salary?*

CS: *Sorry, we don't give out salary information.*

YOU: *All right, but what did you say his title was?*

CS: *Well, we don't really go in for a lot of titles. Jerry had a lot of duties and responsibilities here.*

YOU: *OK, what were some of Jerry's responsibilities?*

CS: *Well, pretty much whatever I assigned him to do. He was really good.*

YOU: *Did Jerry do much in the way of research work for ACME?*

CS: *Oh, sure. Whenever it was needed.*

YOU: *Exactly what sorts of research work?*

CS: *Well, you know, a little sales and marketing, things like that.*

YOU: *How about metallurgical research? Do you do any of that at ACME?*

CS: *I'm not exactly sure what you mean. Could you explain a little more.*

YOU: *Oh, that's OK. I guess you really liked Jerry?*

CS: *Yes, he was great. It's too bad we had to let him go.*

YOU: *And why was that?*

CS: *You mean why did we let him go? Well, I'd really rather not get into that; it wasn't my recommendation.*

YOU: *Well, thank you very much for your time.*

It is probably time to take another look at your list of applicants or go into a much deeper check on Jerry. Why? He does not appear to have had any applicable research experience as required by the job, and his Supervisor didn't even give you the equivalent of name, rank and serial number! Either something is wrong at ACME, with Jerry, or Jerry gave you misinformation.

Exit Interviews

Exit interviews are done when employees leave the company to determine the employee's reason for leaving and to gather impressions of the responsibilities, the supervisor and the organization in general. Depending on the circumstances of the employee's departure, the exit interview may bring out some very candid and often valuable information.

As with the other types of interviews, Supervisors should check first to see what the organization's policy on exit interviews is. Most commonly, exit interviews are the responsibility of Human Resources. The exit interview is most effective when done by an objective third party, such as someone in Human Resources. However, even if you don't do the interview yourself, you may want to share your ideas about what you would like to find out through the interview.

Why should an exit interview be done? As stated above, the information is often candid. This can be a refreshing change from the more common guarded responses in a work situation. Of course, candid is not the same thing as honest in this setting, since employees can also use exit interviews for a vendetta. However, the information gained in an exit interview may assist in the selection of the next employee, or even in development of better job requirements.

The following are a few tips on how to conduct an exit interview.

- It is not a good idea to do exit interviews in the circumstances of lay-offs or a plant closing. This may only add insult to injury and will usually yield very little of positive value.
- Conduct the exit interview a day or two before the employee leaves. The day of departure is usually too emotional, and once the employee leaves she/he will not be likely to come back for an exit interview.
- As with other interviews, conduct the exit interview in a private, relaxed location.
- Use an exit interview form, if available within your organization, or develop your own so that all employees leaving are asked the same questions.
- Areas you may want to explore in the exit interview include: the reason for departure, the new job and salary, advantages of the new position, rating of the present job, supervision, working conditions, safety of the environment, equipment, co-workers, morale within the department and company, advancement opportunities, orientation, training, pay, benefits. Would the employee ever return? Could the departure have been prevented? What did the employee like best/least about the job? Any suggestions for improvement? General comments.
- Document the exit interview and keep the information in a confidential location. This information should not become part of the employee's personnel file, however.

Interviews for Accident Investigations

Conducting accident investigations is often a critical role of the Supervisor, although some organizations retain this function in the Safety Department. Even when a Safety Department has primary responsibility for the investigation, the Supervisor will inevitably be involved. It is, therefore, important to understand the nature of the accident investigation, how it is usually handled, and what Supervisory interviewing skills you will need to apply.

There are two primary purposes of an accident investigation: to determine the cause of the accident (either human, situational or environmental) and to identify steps that should be taken to prevent a similar accident from happening again (Figure 9 – 12).

To determine cause (human, situational or environmental), you have to know the facts. This means you need to collect data, uncover possible indirect accident causes, document facts, particularly those used in compensation and litigation situations, and conduct a lot of interviews. Fact-finding, not fault-finding, is the best way to promote effective safety programs and emphasize management's commitment to safety and health.

To help prevent future accidents, you must have the facts of the present accident. As you determine solutions to prevent the accident from happening again, you also have to look at the cost of the solution or alternative solutions.

Facts in almost any situation seem to be perishable: people forget what they saw. Just as you must confront an employee behavior problem as soon as you see it, you must confront an accident investigation virtually immediately. In deciding when to conduct an accident investigation, the rule of thumb is:

As soon as possible, the sooner the better.

What should be covered? The following is the range of questions that need to be answered in an accident investigation:

1. What was the injured person doing at the time of the accident? Was the person performing the assigned task? Was maintenance a problem? Was the employee assisting another worker or doing his/her own job?
2. Was the injured employee working on an unauthorized task? If so, why?
3. Was the employee qualified to perform the task he/she was doing and was the employee familiar with the process, equipment and machinery?
4. What were the other workers doing at the time of the accident? Who was around? Who saw the accident? What did they see?
5. Was the proper equipment being used for the task at hand (screwdriver instead of can opener to open a paint can, file instead of grinder to remove a burr on a bolt after it was cut, etc.)?
6. Was the injured person following approved company safety/job procedures? Did he/she know the procedures? Was there a Job Safety Analysis (JSA)? Did he/she follow the JSA?

Figure 9 – 12. Accident Investigations

Every accident investigation has two primary purposes:

1. To determine the cause of the accident
2. To identify steps to take to prevent a recurrence.

7. Was the process, operation or task new to the area? The injured employee? Other employees?
8. Was the injured person being supervised? Did the Supervisor intervene? What was the proximity and adequacy of supervision?
9. Did the injured employee receive hazard recognition training prior to the accident? When?
10. What was the location of the accident? What was the physical condition of the area when the accident occurred?
11. What immediate or temporary actions could have prevented the accident or minimized its effect if they had been taken in a timely manner? Why weren't they taken?
12. What long-term or permanent action could have prevented the accident or minimized its effect? What would such preventive measures cost?
13. Had corrective action been recommended in the past but not adopted? Why?

When conducting the accident investigation, you need to interview all victims and witnesses of the accident. You should seek to establish rapport with the people involved by emphasizing the need to prevent recurrence of such accidents. Since liability is always a concern, you need to emphasize the positive and make it clear that you are not on a witch hunt. Witnesses (particularly employees) may be frightened or reluctant to cooperate, especially if they were part of the cause of the accident or if a fellow-employee was at fault. Your professionalism and skill in establishing rapport will be critical to overcoming these barriers.

The following is a sample interview format for conducting and following up an accident investigation interview.

1. Discuss the purpose of the investigation and the interview (fact-finding, not fault-finding).
2. Have the individual relate his or her version of the complete accident with minimal interruptions. If the individual being interviewed is the one who was injured, ask what was being done, where and how it was being done and what happened. If practical, have the injured person or eyewitness explain the sequence of events that occurred at the time of the accident. Being at the scene of the accident makes it easier to relate facts that might otherwise be difficult to explain.
3. Ask questions to clarify or fill in any gaps.
4. Use the same confirmation techniques we use in other forms of interviewing: the interviewer should repeat the facts of the accident to the injured person or eyewitness. Through this review/confirmation process, there will be ample opportunity to correct any misunderstanding that may have occurred and clarify, if necessary, any of the details of the accident.
5. Discuss methods of preventing a recurrence of this particular accident. Ask the individual being interviewed for suggestions aimed at eliminating or reducing the impact of the hazards that caused the accident to happen. By

asking the individual for ideas and discussing them, the interviewer will show sincerity and place emphasis on the fact-finding purpose of the investigation.

6. Immediately summarize the findings of the investigation in writing.

7. Follow up with any witnesses you interviewed in the investigation to let them know the outcome. There is nothing worse than not knowing what happened to suggestions for improvement of safety.

8. Implement any corrective action as a result of the findings of the investigation.

9. Monitor any changes you have implemented.

As you will note, the process of conducting an accident investigation is the same as the process for conducting most interviews. Establish rapport. Get the facts. Confirm the understanding of the facts. Identify solutions. Evaluate solutions. Act on the feasible solutions. Document.

As we will explain in Chapter 11, Discipline in the Workplace, the purpose of discipline is to try and save an employee from termination. Companies invest a great deal of time and effort in training and developing employees. Supervisors, particularly, expend tremendous energy in trying to make sure their employees succeed. Bad safety habits can sometimes become a major liability for an otherwise good employee. If the Supervisor is to help save that employee, some necessary counseling and assistance must be undertaken. Much of this requires effective utilization of interview techniques.

We will illustrate this further with a case study on counseling an employee to improve safety practices.

CASE STUDY: SAVING DANNY

Danny Riley was a well-trained, competent electronic technician. The company required that he use his personal car to travel about the assigned job area as a part of the job of diagnosing and correcting trouble in unmanned microwave stations. Danny was highly regarded by his superiors, had five years experience with the company and was considered one of their best technicians in the Utah region, with a fine potential for advancement to group leader or even district Supervisor.

Since driving a car was a requirement of the job, Clinton Eastbranch, Danny's Supervisor, was concerned about the fact that Danny already had two motor vehicle accidents, approximately one year apart. In the first case, he and Geraldine Cassandra, a fellow employee and electronic technician, were both assigned to temporary patrol duty and were driving from one microwave station to another to check the operational status of the stations. In the Utah region, the microwave stations were spaced about 85 miles apart (depending on topography). Each technician's patrol area covered about six stations. Both employees had been instructed to alternate driving every hour, with the second person assigned solely for safety reasons.

At the time of the accident, Danny and Geraldine were working an 8 p.m. to 4 a.m. shift. Shortly after midnight, Danny went to sleep while driving, and the car ran off the road. Both employees were hospitalized with serious injuries. It later developed that Geraldine was asleep in the back seat at the time and not beside the driver where she could carry out the safety check function. Danny was unable to return to work for over two months.

After the investigation, both employees were informed that another work error of any type could result in discharge. Approximately one year later, Danny had a second motor vehicle accident in connection with work and was suspended pending the investigation and a decision.

In this accident, Danny was driving on a one-way street in the left lane. A slow-moving truck was in the right-hand lane. As Danny approached the truck, he saw a boy on a bicycle cross from the left to the right and disappear from view ahead of the truck. Danny blew his horn and started to pass the truck because he intended to make a right turn into the street at the end of the block in which they were traveling. When his car came even with the front of the truck, the boy on the bicycle, who had decided to recross the street from the right to the left, suddenly reappeared in the car's path and was struck.

Although Clinton Eastbranch, the Supervisor, felt there was no alternative but to discharge Danny, he was completely puzzled by the two accidents, particularly the second one. Danny was a fine employee, one who had no personal injury accidents in his five years of employment with the company. Supervisor Eastbranch, who believed in Danny completely, maintained that there must be some obscure reason. He conducted an accident investigation.

The investigation revealed:

1. Danny should have followed the slow-moving truck, instead of speeding up to cut around it and make a right turn at the end of the block.
2. Danny maintained that, during his driving, he had been thinking out a plan to correct the trouble at the microwave station and get on to the next one. This was consistent with his fine record as an electronics technician.
3. In his youth, Danny had been an enthusiastic automobile racer and had engaged in competitive drag racing.
4. Danny's present hobbies were flying and motor boat racing.
5. When Danny saw the boy disappear ahead of the truck, he instinctively speeded up to pass the truck and make the right turn. This was done without thinking, apparently as a result of his years of race driving.

6. Danny viewed ordinary motor vehicle driving as a sort of humdrum procedure, without using the defensive driving skills he applied in race car driving.

7. In the driving test given following the accident, Danny rated about the same as he did in previous driving checks and in the driver training course required by the company (70 on the scale of 100 placed him as satisfactory but below the expert range.) During the driving test, he talked constantly and turned his head often.

Supervisor Eastbranch recommended and his manager agreed that, as a result of the investigative findings, Danny should not be fired. Instead, it was agreed that Eastbranch would counsel him on achieving a better degree of safety. Supervisor Eastbranch and Danny met, discussed the findings of fact in the case, and mutually decided on the following improvement plan:

1. Danny would continue on his driving assignment.

2. It was made clear that driving was to be his only job while in the driver's seat, and if he found his mind wandering to troubleshooting or anything else, he was to find a safe place to pull off the road, stop, think out his problem and then proceed with his driving.

3. Prior to starting his assignment, Danny and his Supervisor were to sit down and, insofar as possible, plan in advance the work at each trouble location he was assigned.

4. Each of his poor driving habits would be discussed, and the nature of the corrective procedures would be stressed. He would be drilled in the correct procedures.

5. For the first week, Supervisor Eastbranch would ride with Danny on his maintenance trips.

6. The second week, the Supervisor would ride with him every other day. Following that, the supervisory checks were to be less frequent, but unscheduled.

7. The counseling was to be conducted on a planned basis for a full six-month period, with spot-check training and follow-up to be conducted during the second six months.

8. Danny was advised that in the event of another motor vehicle accident, a completely fair evaluation would be made, and he would not be held accountable for things beyond his control. However, if the responsibility for the accident were his, he would be discharged without further recourse.

How would you have handled this case? Do you think an improvement plan should have been developed after the first accident? Why do you think Supervisor Eastbranch waited until after the second accident to develop a plan?

CONCLUSION

Supervisors may need to conduct many different types of interviews over a period of time. The employment interview is a critical one since it sets up the composition of your working team. Each organization has its own policies regarding who conducts what type of interview. Many involve the Supervisor more and more in the interview process, and some make all types of interviews part of the duties of a Supervisor. The hiring interview, the primary focus of this chapter, is one responsibility that is more and more often being assigned as part of the Supervisor's duties. The accident investigation, another critical form of Supervisor interviewing, uses most of the interviewing techniques and skills covered in this chapter.

Preparation and practice are the keys to all forms of effective interviewing. Supervisors must learn to utilize the communication skills learned in Chapter 4 and apply them in interviewing employees or potential employees. Whether in an employment, reference check or exit interview, you must always be aware of cultural differences (Chapter 6) and the employment law implications (Chapter 7) of what you do.

Chapter 10

Performance-Based Supervision: Supervisors and Performance Appraisals

OVERVIEW

This chapter is concerned with performance-based supervision, one of the most important aspects of maintaining and increasing productivity, and thus, a set of skills critical to being a Supervisor. Performance-based supervision, as used here, evolved from the concept of performance-based management. There are significant similarities and differences between the two. Performance-based supervision and performance-based management share the same basic tools—job descriptions, planning, objective setting. They differ in two major ways. Supervisors most often apply these tools to immediate problems and projects, while managers tend to apply them to longer-term problems and projects. More importantly, in a supervisory context, performance-based supervision is almost always primarily concerned with how to deal with human resources, while performance-based management often focuses on objectives for production, marketing, finance, and so on.

Performance-based supervision involves seven main elements:

- Job descriptions—describing your employees' job duties
- Planning—determining how you will reach your work unit goals
- Objectives—setting performance expectations
- Standards—establishing measurements for how performance will be evaluated
- Negotiations—communicating and negotiating objectives, standards and performance with the individual employee
- Evaluations—appraising overall performance of the individual
- Documentation—utilizing effective recordkeeping

All of these elements are codependent and any performance appraisal system is only effective if all of these seven factors work together. In carrying out these key elements, performance-based supervision requires that you use many of the supervisory skills we have discussed in previous chapters. The primary skills you will need to use are:

- Leadership
- Communication
- Talking with the team
- Application of employment law
- Interviewing

Performance-based supervision also sets the groundwork for the skills discussed in upcoming chapters: effective discipline, planning and employee and career development.

This chapter, therefore, will concentrate on the seven key elements where you will need to apply your supervisory skills: job descriptions, planning, objectives, standards, performance appraisal and documentation.

JOB DESCRIPTIONS: DESCRIBING YOUR EMPLOYEE'S RESPONSIBILITIES

If you are going to assess and evaluate individual employee performance, you have to know exactly what it is the employee is supposed to be doing. If you don't at least know that much, you can't go any further. Job descriptions, then, are the first step in the performance-based supervision process. They serve the purpose of describing and clarifying the work activities and responsibilities of each job and, thus, what each person should be doing on each job. Good job descriptions should also establish a chain of command, describing where the occupant of the job reports, whether the occupant of the job has responsibilities for overseeing or coordinating work with others, and so on (Figure 10 – 1).

Components of a Job Description

Job descriptions should state the purposes of the job: why the position exists, its importance to the organization and why the job must be done safely. In addition, job descriptions should include job-related tasks or specific activities stated with action verbs. (See Figure 10 – 2 for a list of action verbs good for writing job descriptions, performance objectives and job safety analyses.) Job descriptions should be reviewed and updated at least annually, preferably at the time of the formal performance appraisal. Of course, if the position changes substantially in the interim, they should be updated accordingly. Job descriptions should also contain the percentage of time the employee will spend on the activities listed.

Figure 10 – 1. What Do Job Descriptions Really Do?

Job descriptions:

1. Describe and clarify work activities and responsibilities of each job

2. Define what each person should be doing on each job

3. Establish a chain of command, up and down

Set-up and Review

Job descriptions, of course, should be put in writing. They should also be made part of an employee's permanent record. That way, when people move from one position to another, the job descriptions will provide an accurate record of their experience. Job descriptions will itemize the skills and qualifications necessary for successfully holding the specific position. Your company will probably have job descriptions and formats established for you and your employees' positions, but you should review them for accuracy. You and the person who holds the job will know how to best describe the position. Share the job description and/or job safety analysis (JSA) with the employee or have him/her help develop it. Review the job description or JSA with the employee and be sure it is understood. Apply the communications skills learned in Chapter 4, Communication: A Two-Way Street. Review the employee's suggested changes with your manager. A typical job description is shown in Figure 10 – 3. In this example, note particularly the following features:

- Action verbs
- Qualifications necessary for the job
- Chain of command
- Estimated percentage of time for each task

Remember, time estimates are *estimates*. They are not to be cast in concrete and should be adjusted with the realities of changing duties and responsibilities.

PLANNING

Planning is an important step in many supervisory responsibilities, such as communication and interviewing. We will discuss this even in more detail in Chapter 12. The planning focus of this chapter is limited to the kind of planning that must be done to make performance-based supervision work. Given that planning point of view, the first step is obvious: you must do some planning. Think how difficult it would be to set an employee's work goals if you, the Supervisor, hadn't first thought out, or planned, what needs to be accomplished. Planning, then, is thinking about what you want to accomplish, how you want to ac-

Figure 10 – 2. Action Verbs for Writing Job Descriptions and Performance Objectives

administer	correspond	examine	obtain	record
advise	counsel	expand	operate	reduce
analyze	create	find	order	refer
arbitrate	criticize	formulate	organize	render
arrange	deliver	identify	oversee	represent
assemble	design	implement	perform	research
assist	detect	improve	plan	restore
audit	determine	increase	prepare	review
build	develop	install	prescribe	route
calculate	devise	institute	present	select
chart	diagnose	instruct	process	sell
collect	direct	interpret	produce	serve
complete	discover	interview	promote	solve
compound	dispense	invent	protect	study
conduct	disprove	lecture	prove	supervise
conserve	distribute	log	provide	supply
consolidate	draw up	maintain	purchase	test
construct	edit	manage	realize	train
consult	eliminate	navigate	receive	translate
control	evaluate	negotiate	recommend	write
coordinate				

Figure 10 – 3. A Typical Job Description

JOB DESCRIPTION

TITLE: Word Processing Operator

PURPOSE: Under supervision, operates a word processing unit for automatic reproduction of letters, reports, agendas or data used on a continuous or repetitive basis.

FUNCTIONS:

1. Operates a WANG word processing unit consisting of a keyboard, video display and printer. Codes machine memory bank for proper formats, typing input onto the system or disks. Time: 30%
2. Selects proper codes to instruct the unit to retrieve information or data required to be reproduced or revised. Time: 20%
3. Prepares listings and related record-keeping of materials contained on information packages, allowing for each search, retrieval and revisions of data. Time: 20%
4. May assist with minor programming of the unit. Time: 5%
5. Performs clerical tasks which may include: proofreading, editing, filing, compiling and collating printed materials. Time: 15%
6. May assist with the training of other Word Processing Assistants. Time: 5%
7. Performs related work as assigned. Time: 5%

REPORTS TO: Word Processing Supervisor

EXPERIENCE: Requires at least one year's experience preferably on a WANG word processing unit.

EDUCATION: Equivalent to a high school education with general business courses. Must be able to type and have knowledge of English usage.

complish it, and laying all this out in a logical format based on as much information as possible.

Tactical Planning

Tactical planning involves a program (plan) of action to advance a purpose or gain an advantage. Take a look at what your department, your work unit, must accomplish in the next week. Based on these immediate needs, your tactical plan will include what can be done in your department to meet those objectives. Tactical planning is short-range and focused on accomplishments that must be achieved immediately. In a military sense, tactical planning is *What do we have to do to take that hill?* In a corporate sense, it is *What has to be done to get the job out this week?* If an organization only engages in tactical planning, it will be responding to immediate needs but may never really address future needs. Thus, it can very easily be caught short by competitors who are looking further ahead. The U.S. automobile industry was guilty of this approach in initially failing to develop high-quality smaller cars. Problems like that are why organizations, including Supervisors and managers, must engage in both tactical and strategic planning.

Strategic Planning

Strategic planning is the process of developing a program (strategy) to meet a purpose or end result. Look at what your organization's goals and objectives are, where the company is going and what role your department plays in moving toward those goals. Rather than the short-range, "must accomplish" standard used in the tactical planning process, strategic planning seeks to look at the big picture. Strategic planning is concerned with what is developing and how the organization will fit into that. In a military sense, strategic planning deals with *How do we win the battle (or war)?* rather than take the hill. In a corporate sense, you will deal with how to make the department or work unit advance toward the department's and organization's goals (Figure 10 – 4). Companies that are obsessed with the big picture and neglect the immediate need also get in a lot of trouble. IBM, for example, was looking at the long-term prospects of mainframe computers while Tandy and Apple got busy tactically positioning themselves in the short-term PC market. It's clear that companies and Supervisors must use both tactical and strategic planning.

Planning and Supervisors

Supervisors, most of the time, deal with tactical planning rather than strategic planning. The pressure of immediate needs is always there and must be dealt with. Nevertheless, there is a real need for Supervisors (and those they supervise) to engage in some strategic planning.

The following example illustrates how both types of planning are essential to maintain safe practices and conditions: Let's assume your crew has to work with some old equipment subject to frequent breakdowns. Let's further assume that the crew has a tendency to take off some of the safety devices to make the frequent repairs easier. It's been a constant battle for you to keep the violations down. You need to engage in both tactical and strategic planning. *Tactically*, you must plan how you will make sure everyone regularly follows good safety procedures even though the frequent breakdowns are a real nuisance. *Strategically*, you need to build a longer-term case to your manager for the newer equipment that will be less subject to breakdowns.

Thus, while meeting immediate needs is important, it also helps if you can figure out how to do things safer and more efficiently for the long run. You are always faced with these two choices: tactical versus strategic. Can you maintain existing equipment, or do you need some new equipment? Is there some additional training that could improve the situation? Should it be done right now? Should it become an ongoing function? If you know a large job is coming up next month, should you schedule vacation time before or after that in order to make sure you have a full complement of people? These all fall into longer-range thinking and, in the context of supervision, are examples of strategic planning. Obviously, we must strike a balance between the two forms of planning. For the Supervisor that balance will usually favor tactical planning because of the immediacy of your job responsibilities. But, don't forget the big picture.

OBJECTIVES

Once you have plans for your department, you can begin developing specific objectives for each employee. When each employee

Figure 10 – 4. Planning

PLANNING IS:

- Thinking about what you want to accomplish
- How you want to accomplish it
- Whether it is logical
- Whether you have enough information

TACTICAL PLANNING IS:

- An immediate gain or advantage
- Short-range objectives
- Getting the job done NOW

STRATEGIC PLANNING IS:

- Steps for meeting long-term goals
- Actions that will take you TOWARD an objective
- The big picture

meets individual objectives, the department will reach its goals and contribute to the organization goals. If individual objectives are not met, you can analyze the original objectives to determine why not.

Advantages

The advantages in setting objectives with and for employees include:

Involvement. Employees can best accomplish what's expected of them if they clearly understand what their objectives are and how they fit into the big picture of the organization. For example, you wouldn't send an employee on a company-paid business trip without first knowing the destination, the reason for going and how to get there. Why, then, would you send an employee to do a job in the shop without being clear about the expected result and why things must be done in a certain way? It's all the same: setting clear objectives.

Job satisfaction. There is satisfaction in setting goals and in achieving them. For example, why do you think so many of us insist on having a list of "Things To Do?" Sure, it helps us to remember what we have to get done, but think how satisfying it is to scratch things off our list. It gives us a sense of accomplishment that, in turn, drives us to set and achieve new goals.

Understanding. Employees can't read your mind. We've heard Supervisors say things like *They know what I expect even if I don't tell them.* This just isn't the case. Employees need to know what the objectives of their work are and what you expect.

Format of Objectives

Although we've said it before, it's worth repeating: Know your organization's objectives and be able to communicate them to your employees and show them how employee objectives impact on the organization objectives.

Well-formatted objectives have the following three characteristics:

1. Objectives should state a result that is to be accomplished. The result should be tangible and easily recognized. It should be derived from the job description, but it is more than just activities.
2. Objectives should be stated in terms that answer what, why, when, who and how. For example, an objective for an employee in the car manufacturing business might be: *Front doors will be installed daily using System 33R2 to meet the organization's production goal of 100 new cars this week.* That simple? YES! But break down the components of that sentence:

What: front doors
Why: to meet the organization's goal of 100 new cars this week
When: daily
Who: the employee for whom this objective is written
How: using System 33R2

3. Objectives should be realistic. They should be attainable but not set so low that they are unmotivating. While realistic, they should involve a stretch (Figure 10 – 5).

Objectives should be set for both short- and long-term plans. Short-term objectives will help gain a sense of accomplishment, but long-term objectives may be necessary to really get the job done.

Task-Related Objectives

Many objectives involve tasks, or clusters of tasks, that must be performed as part of the job. Such task-related objectives should include three components: performance, conditions and standards. They should contain answers to three key questions: how the task is to be performed, where or under what conditions the task will be performed and to what measurable standards the task will be performed (Figure 10 – 6). Let's examine a sample task-related objective to see how it answers these three questions:

In your job as night watchman, one of the assigned duties is to check all the fire extinguishers *on Mondays, during your nightly walk through Building A. You are to record and immediately report any that do not show a reading in the blue safe zone.*

How is the task performed? *You will check all extinguishers and immediately report problems.*

What are the conditions? *You will do this in Building A, each Monday.*

What are the standards? *You will report extinguishers that do not show a blue safe zone reading.*

Task-related objectives, then, can get very, very specific. Having established this, let's move on to the standards set in objectives.

STANDARDS

Standards are measurable indications that objectives are being met in the appropriate way. Standards are the yardstick of the workplace: they show us exactly how our work performance will be measured (or as exactly as possible). Clear standards prevent the "But I didn't know I had to do it that carefully" routine.

Format of Standards

Good standards for performing jobs include four common elements:

1. Standards are task-related, as described previously.
2. Standards should be set in measurable and observable terms. Measurable terms include time frames, quality, quantity and cost tolerances. For example, going back to our door installer, measurable standards could be:
   - 25% of the doors will be installed each quarter (time frames and quantity)
   - Doors will be installed with no greater than a 1% error rate (quality)

Figure 10 – 5. A Performance Objective

A good performance objective should be:
- A result that is to be accomplished
- Easily recognized and clear
- Drawn from the job description
- More than just activities
- Stated in terms that answer What, Why, When, Who, How
- Realistic and attainable, but a stretch

Figure 10 – 6. Task-Related Content of an Objective

To state an objective in task-related terms, ask:

What is the performance expected?

What are the conditions?

What are the standards?

3. Standards should be within the control and accountability of the employee expected to meet them. There is nothing worse than saying you will be measured by a standard outside your control or knowledge. For example, assume you are working as a switchboard operator and you're told that you are expected to meet the company's rate of calls handled per hour. Assume that the first hour you have almost no calls, and the second hour you are swamped. You have no control over the frequency of calls. The standard should specifically spell out the acceptable manner in which calls are handled. Without this information, you're bound to feel anxious and frustrated. Realistically, you can't meet such vague standards.

4. Standards (and objectives for that matter) should be clear, concise and easily understood by the employee expected to meet them.

To develop standards for an objective, you need to look at the task and break it into its three components (performance, condition, standards). Then, look at the standards in light of the separate performance and conditions. Are the standards realistic? Can they be achieved? How can they be achieved safely? Are they what you consider fair and appropriate?

Setting Priorities

Having set some standards, it also becomes apparent that you need to establish some priorities. What things should be done in what order? We mentioned earlier that the amount of time to be spent on each activity should be noted on the Job Description. This is one type of priority. However, we sometimes find that we actually spend more time on the less important activities. Thus, we can't simply say do A, B, and C as if they were equal responsibilities. Usually, they are not. Therefore, we need to establish priorities among our objectives.

Each objective for an employee should also have a priority assigned to it. For example, the automobile plant door installer may only spend 20% of his work time installing doors, but this may be the most important part of the job. The employee who doesn't understand this may continue to do something less important rather than concentrate on accurately installing doors.

Listing objectives in their order of importance is one way of assigning priority. For example, stating that installing doors is the first priority assignment, checking the door panel is the second, polishing the door knobs is the twenty-fifth, and so on, is one way to establish priorities.

The problem with this approach is that it clearly establishes the rank order of the tasks but it does not establish the relative priority of one task to all the other tasks.

Assigning a weight or percentage of importance to each objective is a more common way of establishing priorities. For example, on a 100-point importance scale, installing doors may be 55 points or 55%. Checking the door panel is 25 points, and polishing the knobs is 0.23 points. As you can see, that presents a very different view of the priorities.

NEGOTIATIONS

At the heart of making performance-based supervision work is the idea that the employee and his/her Supervisor mutually agree on just what the objectives are. This is a matter of both communication and negotiation. Negotiating objectives is much like other forms of reaching consensus: you always seek common ground, you try to reach clear understanding. If you cannot reach a consensus, then you must convey to the employee precisely what your decision is and what is expected of her/him. But it works much better if you can negotiate it and reach an agreement.

Getting Ready to Negotiate

Review each of the following key items necessary to have an effective objective-setting negotiation with your employee.

The best way to begin is to tell an employee what the organization's and department's objectives are. Be clear and precise. Get the employee involved in setting objectives, standards and priorities that fit in with those organization and department objectives. Some employees may not be open to negotiation, but for those who are, get the employee's input.

Let the employee know what areas are open for negotiation and what are not. Clearly explain what the objectives and standards are, and show how they apply to the specific job involved. Many employees may not be used to this type of involvement and input, so give them time to think about it before you ask for their specific suggestions. Mutual goals, agreed upon by the Supervisor and the employee, are the best solution. If it doesn't seem to be working, be ready to try again.

Employees also need to think about what resources they will need to meet objectives, including what assistance they will need from you. Give them time to do this.

When you have thought through these steps, it's time for the negotiating meeting with your employee.

Holding the Meetings

Meet with each employee to discuss what he/she has come up with. Give positive feedback on the objectives that are realistic and show the employee's willingness to stretch to attain more and meet quality standards. Compare the objectives with the department's and help bring them into line if necessary. Go over each objective, standard and priority.

Once objectives, standards and priorities have been firmly established, review them one by one with the employee to assure understanding. Allow for questions or comments along the way. Once established, put these objectives in writing and keep them as a part of the permanent record. You and your employee should also sign the objectives document indicating you have discussed it. Make sure the employee gets a copy also.

Set dates for interim feedback on these objectives, so the employee will know that these won't just be put in a drawer, never to be seen again until the annual performance appraisal. Involve the employee in problem solving of potential roadblocks to meeting goals, etc. Get feedback from all employees affected by the plan.

Figure 10 – 7 is an example of a completed employee performance-based set of objectives. The priorities have been established as a percentage, the standards are clear and very precise. In the figure, errors are allowed, but for jobs affecting health or safety, error-free performance is the only acceptable standard.

EVALUATIONS

We have just described a process of developing objectives, setting standards and priorities, and putting it all down in writing. That

Figure 10 – 7. Sample of Objectives, Standards and Priorities

WORD PROCESSING OPERATOR

20% OBJECTIVE:
Word processes weekly production report for Department Supervisor.

STANDARDS:
a. Production reports will be completed by noon on Friday during each business week and submitted to the Supervisor.
b. Reports will have no more than one format error.
c. Reports will have no more than one typographical error.
d. Reports will be stored in the system memory for a minimum of two years.

40% OBJECTIVE:
Quarterly production reports will be word-processed and submitted to the Department Supervisor.

STANDARDS:
a. Reports will be compiled from the quarter's weekly reports.
b. Reports will be submitted by the second week of the month following the end of each quarter.
c. Reports will have no more than one format error.
d. Reports will have no typographical errors.

20% OBJECTIVE:
All department correspondence will be filed on a timely basis.

STANDARDS:
a. A chronological filing system will be developed and maintained.
b. All correspondence will be filed in its proper chronological order within three days of receipt.

20% OBJECTIVE:
The work will be performed safely.

STANDARDS:
a. Work at the terminal will be performed in blocks of time no longer than one hour, with at least 15-minute breaks in between to avoid eye strain.
b. Only one file drawer will be open at a time to avoid tipping of file cabinets.
c. All walkways will be free of electrical cords so that employees cannot trip over cords.

system is at the heart of performance-based supervision. However, the problem with focusing on written appraisals, written objectives, and fixed-date appraisals is that it reinforces a false idea that performance appraisal is something that happens once or twice each year.

The formal part of the appraisal process may occur once or twice per year with written materials. The heart of the process, however, is constant appraisal and communication.

Ongoing Feedback

Ongoing feedback to your employees is a must. If employees did not need feedback, positive and negative, they probably wouldn't need a Supervisor. If you have gone through an appraisal process, then you have just spent a lot of time and energy developing objectives, standards and priorities. Now your employees know what is expected and now you know what to look for in your employees. DON'T, DON'T, DON'T wait for the annual performance appraisal to let them know how they're doing. A couple of our favorite DON'Ts from Supervisors are comments like: *I shouldn't have to praise employees for doing a good job. That's what they're paid for.* or *I shouldn't have to tell him not to leave his tools in the walkway. If he does it one more time, I'll just fire him.* Employees must receive active and continuous communication and this means feedback from Supervisors.

Supervisors' ideas of what workers want from a job differ from what workers actually want, according to a study by Padgett Thompson, a Kansas training firm. A group of Supervisors and a group of workers were given a list of ten factors and asked to rank them in order of importance. The results were as follows:

Table 10 – 1. What Workers Want

Work Factors	Workers' Ranking	Supervisors' Ranking
Appreciation for work finished	1	8
Feeling "in" on things	2	10
Sympathetic understanding of personal problems	3	9
Job security	4	2
Good wages	5	1
Interesting work	6	5
Promotion and growth opportunity	7	3
Management loyalty	8	6
Good working conditions	9	4
Tactful discipline	10	7

One conclusion to be drawn from the lists is that Supervisors overestimate the importance employees attach to financial rewards, and thus probably underuse such nonfinancial rewards as recognition, communication, education, participation and support.

Here are a few tips on providing ongoing performance appraisal feedback (Figure 10 – 8).

- Keep in mind the job description, objectives, standards and priorities that have been set and provide feedback accordingly. Review these with the employee frequently and revise as necessary.
- Observe your employees' performance and give immediate feedback. (Unless you're angry, then wait until you cool down.) If they obstruct a walkway, let them know immediately that they are breaking the safety rules, and you expect the situation to be improved at once.
- Give feedback that is specific. Saying *Good job, Susie* is not specific. Saying, *Susie, you are installing those car doors using the system we expect. Your production is right on schedule by having installed 5 doors this week. That's great.* is specific feedback.
- Be a coach, not a judge. If performance is off track, see what can be done to get it back on track. Don't just say, *You're wrong; fix it.*
- Don't be afraid to give constructive criticism. An employee may not realize he/she is doing something wrong and would appreciate the input. Don't let the employee develop a bad habit before saying anything because it may either create a hazard or be more difficult to correct. *Jim, I see you're not wearing your safety glasses again. I've given you two warnings on this, so I'm filing a written report this time, and I expect no further incidents like this. Now go get your safety glasses, put them on, and wear them when you do this work.*
- Give feedback that is specific to the employee you are addressing. In other words, don't compare one employee to another. Remember how you hated being compared to your siblings or hearing how the star of the class always did things right? Joe doesn't need to know that Jane does the work better than he does. (He probably already knows it, anyway.) He needs to know exactly what you expect of him, not how happy you are with Jane's work.
- Don't criticize an employee in front of others. Injured pride is not a major motivator.
- Don't give feedback to the entire group if it only applies to one or two individuals. Chances are, the people who need to hear it won't be listening or won't think you are talking to them.
- Don't praise just to praise. Undeserved praise does not produce positive results and may instead reinforce unacceptable behavior.
- Keep your feedback consistent with your actions. Don't tell an employee not to leave tools laying about if you do the same thing.

Figure 10 – 8. Tips on Ongoing Feedback

- Remember job descriptions, objectives, standards and priorities.
- Observe your employees' performance and give immediate feedback.
- Give feedback that is specific.
- Be a coach, not a judge.
- Do give constructive criticism.
- Don't compare one employee to another.
- Don't be critical of an employee in front of others.
- Don't give feedback to a group if it only applies to one or two individuals.
- Don't praise just to praise.
- Keep your feedback consistent with your actions.

Interim Reviews

Even though you have just learned how to give immediate and ongoing feedback, it is also a good idea to do it a little more formally from time to time. Your company may well have a specific policy on the minimum number of times you must have a performance appraisal session with your employees. Few companies say anything about the maximum. We recommend that as a minimum, you bring out the objectives and standards document quarterly. Set up a meeting with your employee to discuss progress toward these objectives. Get the employee's opinion on how he/she is doing. Give the employee your specific impressions of how you think he/she is doing, citing examples along the way.

Many Supervisors object to quarterly reviews. However, quarterly reviews are an important opportunity for you and your employee to quickly review where things stand and make frequent, mid-stream adjustments. It is also a time to review plans for the next quarter. Try it.

When you've done this, adjust the objectives and standards, if necessary. You may find that they are not realistic or events have occurred to change priorities.

Formal Performance Evaluations

Make sure you familiarize yourself with whatever appraisal process your organization has. They may have an appraisal form that is to be completed on an annual basis. The form itself is probably the least important aspect of the performance appraisal process, however. The form should only serve as a tool for a discussion between you and your employee. We shudder to think how many Supervisors there are out there who fill out an appraisal form, hand it to the employee without warning, and then send them off to read and sign it on their own. There is no discussion, no chance for the employee to give input, and no chance to develop future plans and objectives. We would agree that type of performance appraisal doesn't work. How could it? We would also suggest that type of Supervisor shouldn't be supervising.

We propose, instead, the following process:

1. Remind the employee that it is time for a performance appraisal. Ask him/her to do a self-appraisal either by completing an appraisal form or making comments on the objectives and standards document. Let the employee know you are sincerely interested in having specific input as to how he/she has done during the past review period.

2. Schedule a meeting with the employee to discuss this information. Plan for the appraisal meeting. Review the objectives and standards and note specific examples of performance relative to these during the review period. Note any particularly outstanding performance as well as areas that need improvement. Keep in mind that you are appraising performance, not personality factors or attitudes.

3. Conduct the appraisal discussion. When you conduct the appraisal discussion you should:

- Present a brief introduction by reviewing the steps that you will go through in the discussion.
- Ask the employee for his/her input, going through the appraisal form or objectives and standards document. Listen attentively and mentally note areas where the employee's assessment and yours agree and/or disagree.

- Present your appraisal verbally. Keep in mind how sensitive the process can be. As Thompson and Dalton stated in 1970, *Performance appraisal touches on one of the most emotionally charged activities in business life—the assessment of a person's contribution and ability. The signals he/she receives about this assessment have a strong impact on his/her self-esteem and subsequent performance.*

- Maintain the employee's self-esteem by emphasizing and praising positive performance. Cite specific examples of how performance has met the objectives and standards.

- In areas that need improvement, try to use the employee's own language if he/she has recognized these needs or shortcomings. If the employee has not noted any areas that need improvement, refer to the standards and objectives and cite specific examples of how they haven't been met. For example, *Tom, we had set the objective that 100 doors would be installed and productivity reports indicate that only 85 doors were installed.* Seek the employees input as to why objectives were not met.

- Again, discuss behavior observations and specific examples of performance. Don't make statements like: *Susie, you are a good employee.* (Why is she a good employee?) *Joe, you have a bad attitude.* (What have you observed that makes you think this?) You might say, *Joe, you left early without permission four out of five days each week.*

- If rankings or grades are used on the performance appraisal form, discuss them at this time. It is best to wait to assign them until you have had a chance to get the employee's input and have discussed performance thoroughly. Give ratings that are consistent with the performance. Don't give someone an outstanding rating if he/she has not met the performance objectives. Probably not all of your employees are outstanding. If they are, your objectives are not set high enough, or you're not being honest.

- Once you and the employee have had an opportunity to review the performance, begin identifying opportunities for improvement or growth. Develop an improvement plan.

- This is also a good opportunity to discuss career goals. Elicit what the employee sees as challenges for the year to come or what type of advancement opportunities he/she is interested in.

- Summarize the discussion. Ask the employee if he/she has any questions or further comments. Set a follow-up meeting to establish objectives, standards and priorities for the upcoming review period.

- If salary is linked to the performance appraisal, discuss it at a later date. The discussion should focus on performance, not pay. The employee should know in advance that you will not be discussing pay at the time of the appraisal. If not, what you have to say about performance may not be heard because the employee is waiting to hear about the pay raise.

DOCUMENTATION

Documentation of performance helps keep the system honest and should be done throughout the entire performance-based supervision system. It also plays an important legal role. We're dealing with a number of things here:

- Assuring communication is accurate, complete and focused where it should be
- Charting progress or lack of it
- Establishing the basis for promotion or possibly dismissal
- Making sure that if it is necessary to fire someone you will be able to do it
- Limiting corporate and personal (yours) liability

What You Must Document

Here's what you must do. There may be additional corporate, state or other record retention requirements and documentation needs.

- Put the job description in writing and keep it as part of the employee's permanent file.
- Put the employee's objectives, standards and priorities in writing. Put any changes that occur as a result of interim reviews in writing. Keep them, probably in your locked desk, but check and follow company policy.
- Make and keep personal notes of feedback that you shared with the employee throughout the review period. State time, date, specific behavior or incident, comments made to the employee and comments made by the employee. Two-way communication and documentation are critical! For example:
  On 1/2/89 I observed Tom's work tools in the walkway. I told him that this was against our safety rules and that they must be moved immediately. I told him I expect that this won't happen again.
  On 2/4/89 I told Tom that I had noticed he had been keeping his tools where they belong and out of the walkway. I told him I appreciated his improvement.

How You Must Document

This may seem time consuming but think how easy it will be to do interim reviews and the performance appraisal if you have all the material on hand. The rule of thumb is simple:

Document at the time of the event to be sure you get the facts of the situation correct.

The performance appraisal will, of course, be in writing. Be sure to address it to the employee, not some mysterious third party. Document the specifics of performance that you presented verbally. Again, stick to behaviors, not subjective opinions or commentaries on the employee's personality. In this role, you are a recorder of behavior and events, not a psychiatrist (Figure 10 – 9).

The objective, here, is that your documentation should be such that if a third party read it, he/she would draw the same conclusions you did in the performance appraisal. Document facts, not opinions.

For example, Supervisors will often address an employee's attitude in a performance appraisal. Attitude is particularly important for employees who deal with the organization's clients or customers. Stating that an employee has a good attitude is only an opinion without supporting documentation. Attitude must be defined and examples must be provided. For instance, a customer service representative can be measured as follows:

- Speaks in pleasant tone of voice
- Answers call within three rings
- Thanks customers for order
- Refers misdirected calls to the appropriate party

LEGAL IMPLICATIONS

In addition to the good solid feedback that the performance-based supervision system provides to employees, it can also make or break an employment law case. The courts are using performance appraisals and related documentation more and more in deciding employment law cases. A good system protects your organization. However, if poorly administered, it can become a liability to the organization and you.

To help you understand this, let's review a few examples of how performance appraisals and related documentation are used in employment law cases.

EXAMPLE 1—FIRING FOR POOR PERFORMANCE

If an employee is fired for poor performance, the courts will look at whether performance standards were set and communicated and whether feedback was given in relation to these standards. If no performance standards were set or communicated and the employee is fired, the court may order the employee reinstated. Or if an employee receives consistently good appraisals and is fired for poor performance, the employee may win the case (possibly including back pay and damages). On the other hand, if performance standards were set and communicated and the employee was given ongoing feedback about how performance standards were not met and eventually was fired as a result, the termination will probably be upheld. *Wrongful termination*, the legal concept involved here, is a rapidly growing area of employment law litigation. Many of the cases that companies lose are lost because of poor documentation.

Figure 10 – 9. Documentation

In this role, you are a recorder of behavior and events, not a psychiatrist. Document facts, not opinions.

EXAMPLE 2—PASSED OVER FOR PROMOTION

Another example is an employee who was passed up for a promotion and claims that it was because of sex or race. Not only will the court look at the performance rating of the individual passed over, but also those of the employee promoted and possibly others who were also passed over for the promotion. Here, as in the first example, documentation makes or breaks the case.

As you can see, the entire performance-based supervision system may be open to review by the courts. This is one important reason to be sure that the system is equitable and that documentation is completed carefully and thoroughly. It is critical. If personnel decisions are made arbitrarily, the organization opens itself to potentially embarrassing and expensive liability. Even if the decisions are not arbitrary, but the proper documentation is missing, the same liability can occur.

CONCLUSION

Performance-based supervision consists of the following key elements: planning, job descriptions, performance objectives, standards, priorities, ongoing feedback, performance appraisals and documentation. Each is a critical part of the process of appraising employee performance on the job.

The performance-based supervision process provides for the development of individual employees to their full potential and assists in the accomplishment of the organization's goals. Effectively used, it also limits the organization's potential legal liability. To achieve these goals requires dedicated Supervisors willing to really apply the procedures discussed in this chapter. It also involves the use of good communication skills and documentation. This process provides for the development of employees, assists in the accomplishment of the organization's goals, and limits an organization's legal liability.

Chapter 11

Discipline in the Workplace

OVERVIEW

Figure 11 – 1. The Ultimate Purpose of Discipline

THE ULTIMATE PURPOSE OF DISCIPLINE IS TO PREVENT TERMINATION.

Disciplining an employee is difficult. Having to confront unacceptable behavior isn't always easy. No one likes to be criticized so it's often not comfortable to have to be the one to tell an employee he or she is not meeting expectations. But as the "Keeper of the Rules," it's your job.

The important thing to keep in mind is that the ultimate purpose of discipline is to prevent termination. Its purpose is humane. It seeks to help the individual and to save time, effort and money by avoiding having to constantly hire and train new people. That may seem like a funny way to look at discipline, but it is the reality of it (Figure 11 – 1).

Effective discipline involves the following steps:

- Catching it early—establishing and communicating rules and performance expectations (discussed in detail in Chapters 4, 5 and 10)
- Doing it now—confronting unacceptable behavior
- Step-by-step—working with progressive discipline
- Dealing with consequences—setting terms for change
- Make a book—documentation again and always and forever (Figure 11 – 2)

CATCHING IT EARLY

Catching inappropriate behavior early, or preventing an opportunity for it, is the best way to reduce your disciplinary work. To do that, people really do need to understand the rules of the workplace and how you expect them to behave and perform their work.

Communicating the Rules

Figure 11 – 2. The Five Key Steps in Effective Discipline

1. Catching the problem early by establishing and communicating rules and performance expectations.
2. Do it now: confront unacceptable behavior.
3. Carry out step-by-step progressive discipline.
4. Deal with consequences: set the terms for change.
5. Make a book: document again, always and forever.

The key to effective discipline begins with the communication of the rules of the workplace and the setting of performance expectations through objectives and standards. You can't discipline or punish employees for breaking a rule they didn't know existed or for not meeting performance objectives if they haven't been told what's expected of them. Yet, time after time, that is precisely what seems to happen in the workplace: things aren't clear. *I didn't know you really meant that* or *I wish you had explained that to me, I just didn't understand that's how you wanted it done* seem like old, familiar complaints. They are. They may also be without foundation, but if you didn't explain clearly and verify the understanding, you'll never know.

Discipline should not be used if an employee has not been trained properly. This is not to say that an employee who is being trained shouldn't be told he/she is doing something wrong. It means that during training the employee should be given an opportunity to make mistakes without being punished.

What Is Discipline?

Disciplinary actions are those steps taken by a Supervisor when an employee fails to meet objectives or standards or fails to follow workplace rules (Figure 11 – 3). Examples of situations that may call for disciplinary actions may include: insubordination, chronic absenteeism or tardiness, violation of safety rules, drinking alcohol or using drugs at the workplace or reporting to work under the influence, being absent without leave, failing to meet performance objectives or standards, fighting on the job, and so on. The list may seem endless, as shown in Figure 11 – 4.

Confronting Discipline Problems

Supervisors may resist confronting a problem. It is easier (in the short run) to look the other way and hope that the problem solves itself. It is also easier to take safety guards off equipment in order to ease repair time. But, since we know that's how you lose arms and legs, we also know that in the long run you're better off keeping the guards in place and doing the job safely and properly. Avoiding the need to confront discipline problems is how you help create employees you have to get rid of. Both good safety and good discipline share the same principle: it is much easier to solve a minor problem early by doing it right than to wait until it has grown into a large problem.

Old habits are hard to break. If you've been used to wishful thinking, always hoping people will correct their behavior on their own, then you will have to break some old habits. If you observe an employee performing a job incorrectly or breaking a rule, address it immediately. If you don't, the employee is probably aware that you saw the behavior and ignored it, the behavior will probably become a habit and it will become even more difficult to change behavior.

Allowing one employee to break the rules will also affect other employees. If an employee doesn't wear safety shoes when required, and this is overlooked by the Supervisor, other employees may stop wearing their safety shoes. Or, if an employee isn't pulling his or her fair share and the Supervisor ignores the problem, this will probably result in poor morale within the workteam. The employees will also lose respect for the Supervisor.

Catching problems early means the Supervisor has to be aware of all employees' performance. It doesn't mean watching over their shoulders every minute of the day, but it does involve being with employees, at least from time to time, as they work. For many Supervisors, that's a given: there's nowhere else to be. However, people with offices (Supervisors, managers, whatever) sometimes seem to lock themselves in an office behind closed

Figure 11 – 3. What Is Discipline?

Disciplinary actions are steps taken by a Supervisor when an employee fails to meet objectives or standards or to follow workplace rules.

Figure 11 – 4. Behavior Calling for Discipline

Insubordination

Chronic absenteeism

Frequent tardiness

Violation of safety rules

Excessive or prolonged coffee breaks

Abusing equipment

Drinking alcohol on the job

Using drugs at the workplace

Reporting to work under the influence of drugs or alcohol

Being absent without leave

Inability to meet performance objectives or standards

Abusive language

Sexual harassment

Racial or ethnic harassment

Overstaying lunch

Too much time in the washroom

Physical abuse

Verbal threats

Fighting on the job

Pushing, shoving

Refusal to follow a direct order

Disparaging the company to customers

Theft from the company

Taking or using company property for personal use

Add a few more:

doors due to the press of paperwork. That doesn't work for managers, and it's a total disaster for Supervisors. If you're not out there, you're not in control. If you're not in control, who is? If you're going to catch problems early, while they are little problems, you must be out where you can see the problems arise.

DOING IT NOW

Obviously, confronting unacceptable behavior is the key to an effective approach to discipline. Let's look at some practical principles involving when, where and how to confront such behavior.

The best time to confront unacceptable behavior is as soon as you see it or notice it. Some behavior is easy to observe, such as a fight or someone not wearing safety gear. Other behavior may be more difficult to pinpoint immediately. For example, an employee may be absent 1 or 2 days, and it may not be a problem. But if an employee accumulates 20 sick days in 2 months, it is a problem. Or, the reduced productivity of a particular employee may not be immediately apparent but should be confronted as soon as it is noticed.

Employees should be disciplined in private. If you don't have an office, find a private place where you cannot be overheard by others. There is nothing more humiliating than being told you are doing something wrong in front of other people, and it may constitute defamation in some situations.

Not all situations that need correction can be handled the same. The first step is becoming familiar with your organization's disciplinary process. This also means knowing what the union contract calls for, if there is one. In some instances, breaking a particular safety rule may carry a specific sanction. For example, if you're a railroader or truck driver and you're caught drinking on the job, the penalty is immediate and specific: dismissal with no recourse. An employee who is not meeting performance expectations may be dealt with in a variety of ways—a talk, suggestions, a schedule of what changes must be made. In most situations, however, progressive discipline is the best method for correcting unacceptable work behavior (Figure 11 – 5).

STEP-BY-STEP

Progressive discipline involves a series of steps or actions designed to improve employee performance. While companies will differ in exactly which steps are to be followed under which circumstances, most use the following steps: coaching or informal counseling; oral warning; written warning; probation; suspension; and, finally, termination or discharge.

The idea is that efforts are made throughout the process to get the employee to correct the unacceptable behavior. Progressive discipline should be used with this philosophy in mind rather than regarding it as a safe way to ultimately terminate someone, although both factors do come into play. As soon as you see behavior that needs improving, you should begin the process of progressive discipline. Failure to do this means you will be stuck

Figure 11 – 5. Confronting Unacceptable Behavior

WHEN: As soon as you see or notice it
WHERE: In private
HOW: Generally, with progressive discipline

with early warning steps to be followed when a problem has already become quite serious. That is not a good position for you or the employee you are disciplining. We'll go over each of these progressive discipline steps (Figure 11 – 6).

Coaching or Informal Counseling

This usually involves a friendly reminder to the employee that a rule is not being observed or a performance objective is not being met. It also includes suggestions from the Supervisor on how to improve. Depending on the seriousness of the situation, you may find it helpful to use coaching several times before moving on to the next step. The use of coaching keeps the process informal and is the best approach to use if you see progress. If you do not see progress (don't kid yourself on this), then it's time to move on.

Oral Warning

This is a more formal notice to the employee that a particular behavior is unacceptable. For example:

Jason, we've talked about the fact that you are tardy two to three times each week. We've reviewed how you can change it by getting up earlier. You seem to make a little progress for a week, and then the next week you slip back. I want you to know that I am putting you on notice: there will be no tardiness next week, and you will need to go at least three weeks with no further tardiness, or I will have to give you a written warning.

The employee is told very clearly he is being warned but given an opportunity to change or improve. Nearly everyone who works for a living understands the difference between an oral and a written warning. Often, an oral warning is enough to cause a change.

Written Warning

You don't simply hand someone a written warning. First, you have a formal discussion with the employee, in private. Following a formal discussion with the employee regarding unacceptable behavior, the discussion is summarized in a written memo to the employee. The memo outlines the behavior that has been observed and how it is expected to change. It should include an action plan that has been developed to help the employee change, including follow-up dates, and the consequences of non-compliance. The employee should sign the memo acknowledging receipt. A copy is then placed in the employee's file. Figure 11 – 7 is an example of a written warning.

It is important to note here one of those clever legal phrases that can be so important. It shows up in the fifth paragraph in the Figure 11 – 7 warning letter:

Your improvement should be immediate, substantial and continuing.

Immediate. Substantial. Continuing. Three strong words. The first word is clear: *now.* The second means big change, major effort, you name it. The third, *continuing,* seems the weakest, but in the long run it is probably the most important. If you include

Figure 11 – 6. Step-by-Step: Progressive Discipline

Coaching/Counseling

Oral Warning(s)

Written Warning(s)

Probation

Suspension

Termination/Discharge

Figure 11 – 7. Written Warning

MEMORANDUM

July 23, 1990
TO: Joe Smith, Lathe Operator
FROM: Carol Lombard, Supervisor
SUBJECT: Disciplinary Warning

Over the past several months, I have noticed you have not been wearing your safety glasses, as the company rules require. You were given a copy of the rules, and I discussed them with you, on January 20, 1990, shortly after you were hired.

I found that you were not wearing your safety glasses on the following dates:
- May 29, 1990
- June 13, 1990
- July 8, 1990
- July 22, 1990

After the first two incidents, I discussed with you, on each occasion, that wearing safety glasses was a rule that must be followed. You showed me that you had your glasses and that they fit properly. You assured me that you would wear your glasses from then forward.

On July 8, 1990, I observed that you again did not have your safety glasses on while operating your lathe. I gave you an oral warning at that time and told you that if this happened again, you would receive a written warning.

On July 22, 1990, I observed that you did not have your safety glasses on, and I am therefore issuing you this written warning. As we discussed again yesterday, wearing safety glasses is an important rule of the workplace. The rule is designed to keep you safe while performing your job duties as a lathe operator. You are foolishly risking your eyesight by persisting in this unacceptable behavior. As you told me yesterday, you have the glasses, they fit properly, you just become careless at times in remembering to put them on at the beginning of the workday. We agreed at yesterday's meeting that you would keep your glasses on top of your work clothes in your locker so that it would be easier to remember to put them on as you prepare for the workday.

In summary, you are expected to wear your safety glasses at all times while performing your job duties as a lathe operator. Your improvement should be immediate, substantial and continuing.

In the future, failure to abide by the rule of wearing your safety glasses while working will result in your immediate suspension from work for three days without pay.

You have demonstrated that you are a skilled welder, and I trust that you will improve in your observance of the work rule we have discussed.

Please sign below to indicate you have received and understood this memo.

Signature: _____
Joe Smith, Employee

cc: Personnel File of Joe Smith

the phrase *Your improvement should be immediate, substantial and continuing* in warning and probation memos, you are, in effect, protecting yourself and the company in the future. That way, if someone improves only while under the gun, you're covered if he/she goes back to old ways after the heat is off. "Continuing" takes away the lifesaver for those employees who know how to play the game of improving only while they are on probation. Use it, it works (Figure 11 – 8).

Under union agreements and even in nonunion companies that have formal grievance processes, there may be several written warnings before you move to the next step. However, assuming the written warning(s) didn't work and the employee hasn't substantially improved, it's time for more serious action.

Probation

Probation is a formal, specified period in which the employee is expected to improve or be suspended or terminated. Some companies use probation instead of suspension and some do the exact reverse. It is not clear which approach is best. The idea involved is that the trauma might just hit home: shape up or ship out. Probation is usually the employee's last chance to improve performance. The terms of probation are discussed with the employee and then documented in writing, similar to the written warning. Certain privileges may be withheld during a probationary period, such as rights to a job transfer or pay increase. This is done because transferring out is a technique frequently used by troublemakers who know how to play the game. Transferring out, under these circumstances, only means you're passing on your problem to another department. The behavior hasn't been corrected, and the company will still have bad effects.

Figure 11 – 8. The Magic Words

YOUR IMPROVEMENT SHOULD BE IMMEDIATE, SUBSTANTIAL AND CONTINUING.

Suspension

Suspension usually follows a severe infraction or the lack of improved behavior after a written warning or probation. The employee is told specifically what behavior has caused the suspension, the duration of the suspension and expectations upon the employee's return. The employee is usually suspended without pay.

Termination or Discharge

The final step should be taken only after careful review that every possible effort has been taken to improve the situation. Documentation must be present that supports this final step. It may be possible to terminate an employee without feeling bad about it. Most Supervisors, even when dealing with worst case types, probably agonize too long and inevitably feel bad about the employee's personal circumstances. Remember, with all these progressive discipline steps you have taken, that employee earned the right to be fired (Figure 11 – 9)!

Don't be afraid to tell the employee why he/she is being terminated. State the precise reasons for the termination. Have the employee consult with the Human Resources Department

Figure 11 – 9. Don't Feel Bad

If you have followed the progressive disciplinary steps, but still must fire an employee, don't feel bad—

THAT EMPLOYEE EARNED THE RIGHT TO BE FIRED!

before leaving to discuss benefits, unemployment, receipt of last paycheck, insurance, etc. Document the termination.

DEALING WITH CONSEQUENCES

Having reviewed the process of progressive discipline, we're going to take another look at how to handle unacceptable workplace behavior. Now, we will focus on setting the terms for change, reviewing the steps to be taken in the event of unacceptable behavior, trying to change that behavior and dealing with the consequences.

Investigate

Getting the facts is critical. If you see something, fine. Confront it. If you only see the results of unacceptable behavior (lost time, production down, stories about what happened) you have to investigate. A few simple rules of workplace investigation:

- Don't act upon gossip or hearsay.
- Pinpoint the problem by assessing whether the employee has been trained properly.
- Interview other employees involved, if appropriate.
- Review the employee's work history and performance record to determine if this problem has surfaced before or if this is a long-term employee with a good work record (Figure 11 – 10).

Meeting with the Employee

Having collected as many facts as possible, it's time to sit down with the employee. Plan the meeting. Have the facts of the investigation on hand. Understand what your objectives are.

Meet with the employee in private and in an unemotional setting. Don't call the employee in when you are angry and tempted to yell and scream instead of discuss. Open the meeting by stating the purpose and describe what you see as the problem. Be specific and straightforward. Maintain the employee's self-esteem by stating the behavior that is a problem, not that the employee is a problem. Don't be accusatory or demeaning. Put the problem in perspective for the employee. Find something to praise the employee for, if possible. For example,

> *Joe, you have had an above average attendance record for the past five years that you have been with us. I am concerned, however, that you have been absent 10 times in the past two months.*

Ask the employee to problem solve the situation with you. Remember that the employee has a right to be listened to and treated with respect. Determine if there are any mitigating factors involved. There may be underlying problems that the Supervisor was not aware of, such as the work assignment itself, family problems, illness or a conflict with co-workers.

Ask the employee what he/she will do to solve the problem and listen to what he/she has to say. If the employee becomes

Figure 11 – 10. Investigate: Get the Facts

- Don't act upon gossip or hearsay
- Pinpoint the problem
- Interview other involved employees
- Review the employee's work history and performance record:

  Has this problem surfaced before?

  Is this a new problem?

  Is this is a long-term employee with a good work record?

angry, the Supervisor must stay calm. Don't start arguing with the employee. Expect that the employee may need to let off some steam. If the employee becomes too emotional at this point, suspend the conversation and let him/her know you will reconvene when the employee has had a chance to calm down.

Using the employee's suggestions, if any, spell out the specific behavior that you expect in the future. Offer any assistance or suggestions you may have. Discipline can reinforce negative behavior, so focus on the desired behavior. The Supervisor should be part of the solution. Determine if there are any barriers for the employee, and offer solutions such as further training or the company's Employee Assistance Program if personal problems are an issue. Continue to maintain the employee's self-esteem by expressing your confidence in the employee's ability to change or improve.

The disciplinary process should not be suspended just because an employee may be experiencing personal problems. The Supervisor should be supportive and offer whatever assistance the company may have, but poor performance must always be addressed. Often, the need for job security provides the type of motivation an employee may need to seek help or to address a personal problem. If overlooked, the performance problem and the personal problem may only get worse.

Set an achievement timetable. Set follow-up dates to meet with the employee to discuss progress. Some problems may require immediate improvement, such as fighting on the job or wearing safety equipment. Other situations, such as tardiness due to family problems, may require more time for the employee to show improvement. Make sure you follow up when you say you will.

Most importantly, explain the consequences of noncompliance. Although you want to focus on the positive, the employee has the right to know what the next step will be in the disciplinary process if the problem doesn't improve. Explain any positive consequences of improved performance. For example, if an employee's pay raise is being suspended because he/she is being placed on probation, explain that he/she may be eligible for a raise if the performance improves and is maintained at an acceptable level.

Finally, summarize the conversation and plan of action. At that point, the meeting is over. Try to end it with a positive note, a hand shake, an expression of your confidence in the employee's ability to change (Figure 11 – 11).

Follow Through

Comply with any follow-up dates that have been set. Don't hesitate to take the next disciplinary step if performance doesn't improve. Try a new plan of action, however, if the first one didn't work. If performance does improve, make sure you tell the employee that you have noticed the improvement and praise accordingly. If the employee's file has been documented regarding the problem, put a notice in the file summarizing the improvement and let the employee know you are doing this. DON'T IGNORE IMPROVED PERFORMANCE!

Figure 11 – 11. Holding the Discipline Meeting

- State the purpose.
- Describe the problem.
- Allow the employee to respond.
- Ask the employee how he/she would solve the problem.
- Don't start arguing with the employee.
- Realize the employee may need to let off steam.
- Use the employee's suggestions to help spell out the specific behavior you expect in the future.
- Offer any assistance or suggestions you may have.
- Focus on the desired behavior.
- The Supervisor should be part of the solution, by offering help and guidance.
- Set an achievement timetable.
- Explain the consequences of noncompliance.
- Summarize the conversation and plan of action.
- Try to end it with a positive note. Don't say "Have a nice day."

MAKE A BOOK

We go through a lot of steps trying to save an employee from termination. But when it can't be done, it can't be done. Even charitable organizations cannot continue to employ people who will not do the work the way it must be done. Organizations, particularly for-profit corporations, hire people to do a job. If they can't do that job, they usually try to train them. But if they just won't do the job, then they should be discharged.

If you went through all of the steps of progressive discipline, trying to help an employee correct unacceptable behavior, and finally determined that it just wouldn't work, how would you feel if the company lawyer said *Sorry, you just can't fire that employee because you have no documentation.*

Just as documentation was critical in the performance-based supervision process, documentation is essential in the disciplinary process. As Supervisor, as the person who deals with the employees of the company, you may be called upon to defend your actions in court or in a labor arbitration hearing. Your documentation should be clear and state the facts of the case, not opinions. Memory alone is seldom a reliable tool for this type of detail. Written records are essential.

Documentation should include the dates of the incidents that led to the disciplinary step being taken, a summary of the incident(s), conversations with the employee, action plans developed and implemented (such as training, coaching, etc.) follow-up steps, and any formal written correspondence, such as warning memos or probation memos. Different companies will have different policies as to where these documents should be kept. Check company policy on this and follow it rigidly (Figure 11 – 12).

DISCIPLINARY TIPS

Discipline should be viewed as a process to help an employee improve performance, NOT as a necessary routine that has to be followed in order to get rid of a problem employee. Yes, it does serve both purposes. But if it comes to that, companies could devise a much faster and less humane system just to get rid of people. Some authoritarian companies do just that. To avoid such circumstances, here are a few tips you should follow.

- Be consistent in your application of the discipline process. Don't warn one employee about excessive tardiness while suspending another employee with the same track record. Inconsistent application of the rules may be interpreted as discriminatory under the law and is simply not a good supervisory practice because it also leads to charges of favoritism.
- Be aware of what your company's disciplinary policies are because you must not only be consistent within your department, but you must be consistent with company policy and practice.
- Don't block your employees from using the formal grievance procedure or from taking the case to a higher

Figure 11 – 12. Why Document?

Good documentation keeps everyone out of the courthouse.

court, such as your boss. They, like you, are entitled to a fair process.

- Consult with the appropriate parties when invoking the disciplinary process. Keep your boss informed of the situation. Seek advice from the organization's Human Resources Department.

- If you're supervising under the terms of a union agreement, remember that agreement is a mutually approved contract reached between your employer and the union. Know and follow the grievance and discipline steps set out in the agreement.

- It is never too late to begin addressing a performance problem or violation of work rules. Even if you have let a problem go unchecked, begin coaching the employee and helping him/her to get back on track. It may take more time for the situation to improve, but now is better than later (or never) for beginning to turn the problem around.

- As always, set a good example. You are the role model. You won't have much credibility, or success, if you're disciplining an employee for something you are doing wrong yourself. If you expect your employees to follow the safety rules, be sure that you follow them. If you don't want your employee to be tardy, make sure that you get to work on time. Do as I say and not as I do just doesn't cut it in the workplace (Figure 11 – 13).

The following case study raises some questions about how you pull the facts together in a discipline matter.

CASE STUDY: GATHERING THE FACTS

Supervisor Bryan Miller had a problem on his hands. The performance of one of his best employees, Cathy Smith, had been slipping the past few months. At his wits end, he went to Diane Drake, the company's Employee Relations specialist. The following conversation took place.

BRYAN: *Diane, I hate to bother you, but I really have a problem on my hands. I hate to say it, but it looks as if I'm going to have fire Cathy Smith, who has been one of my best employees in the past.*

DIANE: *Really, Bryan? If I recall, Cathy's Performance Review was really good.*

BRYAN: *Yes, I know! But for the last two months she's been coming in late, taking long lunch hours and making a lot of mistakes on the job. I've mentioned all of this to her a couple of times, but her performance just hasn't gotten any better. I don't know what else I can do.*

DIANE: *Well, Bryan, what steps have you taken to help Cathy improve? For instance, are her performance expectations clear, defined and documented?*

BRYAN: *I think so. For the most part, she's well aware of her performance expectations. I added a few responsibilities to her job a few months ago, but I haven't really had the time to sit down and spell out the objectives and standards on these. Anyway, Cathy is one of those people who always did well, so I figured she'd just pick up on the new duties and do a good job, like she's always done. Even so, she knows the work rules on getting to work on time and the 45-minute lunch period.*

DIANE: Bryan, I'm sure you are right about Cathy knowing the rules. But my question is: Has she really been trained for these new responsibilities? Does she really know what you expect if you haven't specifically covered it with her?

BRYAN: I think I'm getting the drift of this. I see what you mean, Diane. Maybe Cathy doesn't know, and maybe she does. But unless I really cover it with her, I can't be sure, and I'm not really being fair to her either! If I do have to fire her, I guess I had better get this part straight. I need to do a little more work before I think about firing Cathy.

What steps would you take if you were Bryan? Would you fire Cathy? Would you go back and outline Cathy's performance expectations and spend more time training her on the added job responsibilities? Would you document the steps you take? Cathy may be coming in late and taking long lunch hours because she is uncomfortable with her job responsibilities. Maybe, maybe not. Remember, before you can know the answer to that, you have to do your job as a Supervisor—step-by-step and with documentation.

CONCLUSION

When it comes to employee discipline, remember to:

- Communicate your expectations and the rules up front.
- Catch and confront unacceptable behavior early.
- Use a progressive disciplinary process.
- Meet with the employee to identify the problem, seek input and develop a plan of action.
- Follow through by praising improved performance and noting areas that still need improvement.
- Document the facts and behavior.
- Be consistent.
- Be careful.
- Above all, remember that you are trying to help the employee.

Chapter 12

Planning, Decision-Making and Overcoming Resistance to Change

OVERVIEW

We have discussed planning in several chapters of this book. Those chapters covered planning hiring interviews (Chapter 9), communications with your employees (Chapter 4), performance appraisal discussions (Chapter 10) and disciplinary meetings (Chapter 11). In this chapter, we will examine planning in more detail as it applies to the everyday work of supervision. In addition, we will explore the decision-making process that can be used in planning as well as most other aspects of supervision. Finally, we will deal with what has become an increasingly important problem in the workplace: overcoming resistance to change. All of these, of course, relate to planning and the Supervisor. Specifically, we will focus on:

- The importance of planning
- The process of planning
- Decision-making and problem-solving
- Overcoming resistance to change

THE IMPORTANCE OF PLANNING

Planning is an integral part of any organization and of your job as a Supervisor. In Chapter 10, Performance-Based Supervision, we reviewed the importance of tactical and strategic planning, pointing out that the organization or Supervisor that engages in no planning or only one form of planning will eventually be caught short. The Supervisor who engages in planning has a far better chance of succeeding.

Your first response to "Planning" may well be *I don't have time to plan* (Figure 12 – 1)! Most first-time Supervisors are used to doing something, some observable activity or task, to get work accomplished. Planning may seem very alien. The fact is, though, you do already plan. At a minimal level, you plan what will be done that day by thinking about what there is to do. You plan who will do it by reviewing what workers are there that day and what they can do best. Whether you think you have the time or not, you are caught up in planning right now!

Figure 12 – 1. Not Now, You Don't

The objective of this chapter is to make your planning more effective, more organized and a little smarter. It may take some practice, but taking the time to sit down and do some thorough planning can save you, your employees and your company a lot of wasted time, energy and money, while you also get the job done more safely. That's doing things a little smarter.

Benefits of Planning

Planning helps you avoid drifting aimlessly from one project to the next, or worse yet, leaping from one fire to another. Planning also helps you avoid supervision by crisis.

Do you ever feel like you go from one crisis to the next or from one problem to the next? Planning can help put an end to this stressful situation. You may solve problems, make decisions and jump to conclusions, but how much time do you spend planning? If you did more planning, you could avoid having to make decisions without all of the facts, or solving problems that never should have occurred in the first place.

You already know that quick fixes don't usually solve a problem long term, anyway. Planning can help avoid problems such as safety hazards and having to create quick fixes. Planning will not eliminate all crises, but it can help reduce accidents and costly mistakes.

Some of the benefits of planning include (Figure 12 – 2):

- Defining how employees' time and energy will be used and how department materials and equipment will be allo-

Figure 12 – 2. Planning Benefits the Supervisor

- Plans help define how to best use employee time and energy.

- Plans define efficient use of department materials and equipment.

- Plans allow effective scheduling of vacation time.

- Plans prevent running short on inventory.

- Plans give a Supervisor greater control over work.

- Plans set priorities.

- Plans allow you to measure results.

- Plans aid in communication.

cated. With a little planning, you can schedule vacations around peak periods and avoid being caught short on inventory.

- Having greater control over the direction and progress of a work assignment or project. You begin to set the priorities for a change.
- Measuring results against established goals and objectives. If your manager thinks your crew didn't move fast enough on the last assignment, planning will provide you with clear measures to show just how well you did or didn't do.
- Providing communication, guidance, influence and leadership. When you plan, people can understand what you want, where you're going.
- Discovering potential safety hazards. You can see and eliminate hazardous conditions or procedures before they lead to an accident.

Effects of Not Planning

Lack of planning can cause many negative results, such as (Figure 12 – 3):

- Increased accidents and safety violations. *I know I was supposed to read those HazCom notices. I just didn't find the time!*
- Lower productivity. *Well, we would have gotten it out yesterday but half my crew were off on vacation.*
- Lower quality of work. *Yes, I know there were quite a few rejects in that batch. I'm really sorry, but we were trying to move both jobs at once, and things just got a little out of hand.*
- Low morale among employees resulting from misunderstanding and confusion. *I told you people this job had to be out tonight. It's just too bad it's taken 12 hours. If you knew your jobs, we could have gotten it done on time.*
- Pressure from your boss due to all of the above. *You did what? Well you better just go right back and do it the right way this time!*
- Fewer opportunities for your promotion or job security. *Promotion? After the mess you've made? You have to be kidding!*

What to Plan

Almost every aspect of a Supervisor's responsibilities can benefit from proper planning. Planning is particularly useful in the following areas:

- Employee management, such as training, recruiting and safety. Setting aside the time for regular safety meetings is a critical part of maintaining safety. You have to plan to be safe.
- Use of facilities, materials and equipment. Who gets to use what, when, is the source of many workplace disputes. Plan the use.

Figure 12 – 3. The Results of Not Planning

Failure to plan can cause:

- Increased accidents
- Safety violations
- Lower productivity
- Reduced quality
- Low morale
- Pressure from your boss
- Fewer opportunities for promotion
- Poor job security

Figure 12 – 4. What Is Planning?

"Planning is to devise or project a method of action." Webster

"My interest is in the future, because I am going to spend the rest of my life there." Charles F. Kettering

- Frequently recurring or critical jobs, projects or assignments. If something happens regularly, it is a prime target for planning. Build around it.
- Work schedules. You know that.
- Productivity and quality control. It doesn't happen by magic. It can happen with careful planning.
- Cost reduction and control. Firefighting is an expensive business. Overtime because of a lack of planning is money that could have been spent on new, more efficient equipment.
- Conservation of time, energy, power or utilities. Smart use of power takes planning.
- Communication with employees, your boss, other management personnel and customers. Regular meetings, reports, good customer services don't just happen. You have to plan.
- Career planning and self-improvement. Without planning, forget it. With planning, you can determine the best course of action to advance yourself.
- Smelling the roses. Now we're talking major planning. With good planning, you'll actually have more time for enjoying life.

So what is planning? As Webster put it, *Planning is to devise or project a method of action.* As the industrialist inventor Charles F. Kettering put it: *My interest is in the future, because I am going to spend the rest of my life there.* Somewhere between those two is a plan (Figure 12 – 4).

THE PROCESS OF PLANNING

So how do we go about planning? Planning is a process. The process is made up of a series of steps, all of which are critical to the completion of planning (Figure 12 – 5).

Six Steps of Planning

Time. The first step in planning is to realize it will take time. In the long run, it will also save time. But at the start, it seems to just gobble up time. Supervisors should give themselves some quality time to plan. That means they should find time when they won't be interrupted or disturbed. All too often, that may mean some at-home time. Planning takes some time to reflect on the past, the present and the future.

Knowledge. The second step in the process of planning is knowing and understanding the company's objectives. The objectives may be formalized in strategic plans (long-range company plans that define what product/service will be provided and to whom), mission statements or other planning documents specific to your organization. You may wish to clarify the company's objectives with your boss so that you fully understand not only the what, but the why, of these objectives. Finding out from your boss or senior management how these objectives were determined may be a critical step in the planning process and everything you do as a Supervisor. This information is not only important for you

Figure 12 – 5. Necessary Steps for Staff Planning

- Determine the kinds of skills and talents needed to meet short-term and long-range objectives.
- Compare skills and talents of current employees against those needed.
- Identify number and kind of skills and talents lacking.
- Develop a plan outlining how and when needed skills/talents will be acquired.
- Identify number and kinds of skills that can be developed through training or education.
- Identify recruitment needs.
- As with all plans, review periodically.

as you go through the planning process, but you should also be able to communicate this to your employees.

Find out the company's short-term plans as well as the strategic or long-term plans. The organization's plans (and ultimately, your plans) provide direction for resources. They are like a unifying framework in which all the work will be done. These plans help to reveal future opportunities and control of resources, and they prevent piecemeal decisions from being made. Basically, these plans guide company resources to the desired results.

Your department or work-unit objectives are actually based on these broad company objectives. Understand what objectives your department is expected to meet. Plans must be developed to meet these objectives. Some of these plans may address day-to-day operations, such as scheduling or materials distribution. Other plans may involve more initiative on the part of the Supervisor, such as how to meet an objective more cost-effectively than has been done in the past.

Just as with the company objectives, Supervisors must understand and be able to communicate to their employees the what and why of these objectives. When you have completed these departmental objectives, they will become tools to help you figure out what financial, material and human resources you will need to meet the objectives. Determine the priorities that are set by these objectives and plan accordingly.

Choices. The third step is to determine the choices you have for meeting the established objectives. Begin by looking at one objective or project that is to be carried out or completed. Ask yourself a few questions:

- WHAT:
  What is the objective or result that is to be achieved?
  What is a priority?
  What pitfalls may occur throughout?
  What may cause trouble and what won't?
  What do we need and what don't we need?
  What is and isn't possible?
  What is the current and future status?
- WHERE:
  Where is the plan to be carried out?
  Where are the products or services delivered?
- WHEN:
  When is the project/objective to be completed?
  Are there time frames or checkpoints needed along the way?
- HOW:
  By what methods or procedures will the objective be met?
  Brainstorm by coming up with as many "hows" as possible, even if they don't at first seem likely or practical.
  How will job or task sequencing and assignments be made?
- WHO:
  Should an individual or group be involved in carrying out the plan?
  Who should have authority and control?
  Who should be involved in helping to develop this plan?

(The sooner your employees are involved, the more they will buy into your plans.)

Who should be informed about your plan and its related time frames?

Answering these questions should give you many choices for meeting the objective. You may want to test some options to determine which are the best. Make sure your choice will best meet the objective or goal.

Documentation. Prepare a written plan, including an action plan (how you will get things moving). Action plans or supporting plans may have to be developed in addition to the overall plan to cover some of the details or options. Supporting plans may have to be developed if additional equipment, materials or employees will be needed to carry out the plan.

For example, assume you begin to look at the need for new people. What kind of supporting plans would you need to develop? If it involves a lot of people or new skills, you have one set of problems. If it requires some special training, perhaps you have another. Figure 12 – 6 shows the range of questions you might need to deal with when it comes to "people planning."

Your written plan should include the objective, the steps to be taken to meet the objective, dates when steps will be taken, materials or equipment that will be utilized, employees who will be responsible for or will be carrying out the plan, priorities and a completion date. Figure 12 – 7 is an example of a Supervisor's program plan. Notice the objective, logical sequencing of steps and identification of equipment needs.

Implementation. The fifth step is to implement the plan. If employees haven't been involved in the planning process, you need to communicate the plan to them. Because such communication may require a lot of meetings, it's best to bring employees into the planning process at an early date. No surprises and giving employees a sense of ownership can promote acceptance of a plan better than anything.

Review. The sixth and final step in the planning process is to review what happens and how the plan is carried out. Keep the plan on track by checking the dates and accomplishments along the way. Check to see that the objective has been met. Adjust the plan, if necessary, to make sure the results are what is needed.

Figure 12 – 6. The Planning Process

- Make a time commitment.

- Know your company.

- Identify choices.
  WHAT
  WHERE
  WHEN
  HOW
  WHO

- Put it in writing.
  Action plan
  Supporting plans

- Implement the plan.

- Review the plan.

DECISION-MAKING AND PROBLEM-SOLVING

Planning involves decision-making. Supervisors make numerous decisions every day. Some decisions are very routine, like deciding whether or not to put on a hard hat if your workplace is a construction site. This decision should be an easy and habitual one. Other decisions are more difficult. Decisions that involve solving a problem can be more complex. Making decisions involves choosing among the available alternatives to solve a problem, complete a project or meet an objective. Decision-making is key to the planning process and to making sense out of your everyday responsibilities as a Supervisor.

Figure 12 – 7. Program Plan

Project: Door Installation Process

Date: 2/24/90

Objective: To decrease the amount of production time currently used in installing doors on cars using System 33R2

Steps	Completion Date	Employee Responsible	Equipment
1. Meet with the boss to clarify purpose of proposed change.	3/1/90	Supervisor	
2. Schedule meeting with workgroup to introduce needed change and solicit ideas for improvements.	3/5/90	Supervisor	
a. Reserve conference room.	3/4/90	Secretary	Room/Flip Chart
b. Prepare agenda.	3/5/90	Supervisor/ Secretary	
c. Notify workgroup of meeting.	3/6/90	Supervisor	
d. Distribute agenda.	3/6/90	Secretary	
3. Hold meeting.	3/10/90	Supervisor	
a. Explain needed change.			
b. Answer questions.			
c. Solicit suggestions to improve installation system.			
d. Decide on one suggestion to try to implement.		Supervisor/ Workgroup	
4. Implement change. (Prepare new program plan.)	4/1/90	Supervisor/ Workgroup	
5. Review new process and make any necessary adjustments.	Ongoing	Supervisor/ Workgroup	

Popularity of Decisions

Some of your decisions may not be popular with your employees. A decision to discipline an employee is necessary but will not be popular with the employee involved. Having to lay off some of your workforce will not be a popular decision but may be necessary at some point. Deciding that everyone has to stay late on the eve of a holiday to finish a project is not going to be popular. Being a Supervisor may mean you will need to make unpopular decisions sometimes. Worrying about popularity and making the wrong decisions because of those concerns leads to disaster. Worrying about popularity and failing to make difficult decisions undermines your supervisory role. It is important for you to make decisions, popular or not. Identify yourself with your decisions and take responsibility for them. Ultimately, you will hold everyone's respect for doing that.

Quality of Decisions

Keep in mind that it is the quality of the decisions you make that counts, not the quantity. We all have experiences with people who have poor decision-making skills. They may procrastinate in making decisions, putting off making a decision until they are confronted with having to make a decision on the spur of the moment. At that point, they usually make the wrong decision. Or they may vacillate, agonizing like Solomon, making a decision, reversing it, then reversing it again. Or they may make impulsive decisions, without regard to the facts of the situation, the feelings of the people affected by the decision, or the cost of the decision.

If you have ever worked with or for people who have these decision-making problems, you know how frustrating it can be! Ultimately, all of them make decisions, but the quality of their decisions is the problem.

Decision-Making Process

If you follow the steps of the decision-making process, you will move yourself toward quality decision-making. These steps to making sound decisions or solving problems are set out below (Figure 12 – 8). They're really simple, but important to follow. They can be used, as illustrated, for different kinds of decisions and are not limited to planning.

1. Realize that you have a situation that requires making a decision or solving a problem. Be aware of symptoms that indicate you may have a decision to make. For example, you begin to notice alcohol on the breath of one of your employees every morning. If the employee works with dangerous machinery, you have a safety problem on your hands and some decisions to make. You need to go through the decision-making process to determine what to do about it.
2. Define the problem and the decision that needs to be made. In this case, the problem is an employee who is reporting to work with the smell of alcohol on his/her breath. The decision that needs to be made is what you are going to do about it. The impulsive decision-maker would at this point decide to discipline the employee for reporting to work under the influence. We suggest you go on to the next step.
3. Analyze the situation.
   - Gather the facts. Is the employee coming in late? Is the employee stumbling, or does he/she have slurred speech? Has his/her performance suffered recently?
   - Determine possible causes. The employee is drinking before he/she comes to work. The employee uses mouthwash that has a lot of alcohol in it.
   - Develop alternative solutions by brainstorming. Call the employee into your office and tell him/her what you have noticed and ask for his/her comment. Look for a bottle in his/her locker (most companies retain the right to

Figure 12 – 8. The Decision-Making Process

1. Realize you have a situation requiring a decision.

2. Define the problem and the decision that needs to be made.

3. Analyze the situation.

4. Make a decision by choosing the best alternative.

5. Remember KISS.

6. Implement your decision.

examine lockers). Ask co-workers if they have noticed anything strange about the employee. Fire the employee.

- Consider the pros, cons and consequences of each alternative. The "pro" of calling the employee into your office is giving him/her an opportunity to explain the situation; the "con" is that he/she may deny that anything is wrong. The "pro" of looking in the locker is that you may find a bottle; the "con" is that you are invading the employee's privacy and this may be illegal in your state. The "pro" of asking co-workers is that they may substantiate your suspicion; the "con" is that this will cause poor morale among the workteam. The "pro" of firing the employee is that you have a quick fix; the "con" is a possible wrongful discharge suit and the loss of a valuable employee.

4. Make a decision by choosing the best alternative. You should gather all the facts relevant to the situation and then call the employee into your office for a talk. It is usually a good rule of thumb to not make a decision without having all, or at least most, of the facts at your disposal.

5. Remember KISS. Your decisions should also be simple, realistic and as inexpensive as possible.

6. Implement your decision. Acting upon your decisions may start the process all over again. For example, if the employee admits that he/she has been coming to work under the influence, you then have a decision to make about what to do about it. If he/she denies it, you have a whole different set of decisions to make.

- Determine how much latitude you have in your organization to make decisions. For example, some companies require that disciplinary decisions be made only with the assistance of your manager and/or Human Resources. You may also have limits in making decisions that will cost the company money, such as purchasing new equipment or firing people.
- Get your employees involved in making decisions, when possible. Delegate the decision, if appropriate.
- Explain your decisions to the people affected by them. Failure to explain your decisions may cause anger, suspicion, misunderstanding, hostility, lack of confidence in your supervisory abilities, frustration, mistrust and lack of support of your decision.
- Make decisions as quickly as possible. This doesn't mean being impulsive, but delays only make problems grow worse. Set reasonable timetables for making decisions, and set limits if fact-gathering drags on and on.
- Keep an objective in mind throughout the process.
- Don't let your emotions or personal preferences get in the way of making decisions.
- Realize that making decisions often involves some risk. Risk-taking can't be avoided if you are a good Supervisor. If you go through the decision-making process that we have described, you have done the best you can to reach a decision. Recognize that you can't do more than

Figure 12 – 9. Tips on Decision-Making

- Determine how much decision-making authority you have.
- Get your employees involved in making decisions.
- Delegate the decision if appropriate.
- Explain your decisions to the people affected.
- Failure to explain your decisions may cause anger.
- Make decisions as quickly as possible.
- Don't make impulsive decisions.
- Usually, delays only make problems grow worse.
- Set reasonable timetables for making decisions.
- Set limits for fact-gathering.
- Keep an objective in mind throughout the process.
- Don't let your emotions get in the way.
- Making decisions often involves some risk.
- Risk-taking can't be avoided.
- Making a decision often involves implementing a change.
- With change may come resistance.

that. Gathering the facts, considering alternatives and not rushing into something will help lessen the risk involved.

- Making a decision often involves implementing a change. With change may come resistance (Figure 12 – 9).

OVERCOMING RESISTANCE TO CHANGE

Change is an inevitable part of life. Change is particularly inevitable in the business world if an organization is going to survive. There is new technology, new customer demands, new employee concerns and expectations, foreign competition and domestic competition. There are also mergers, acquisitions, downsizing, takeovers, sell-offs, spin-offs and all of the other combinations of corporate restructuring that have become a part of everyday life. If organizations don't change in the midst of all of these factors, chances are they won't be around much longer. Maintaining the status quo just doesn't work. As a Supervisor, you must learn to introduce change to procedures, machinery and your staff.

If you have not experienced a corporate restructuring, you're lucky, but your luck may change. As a Supervisor, you will also be responsible for figuring out how to live with and defend changes that you and your employees may not like. This is not a pleasant prospect, but one faced by many Supervisors, managers and employees. People are almost always upset when their workplace is changed, but they do learn to adjust, and many make the adjustment quickly and effectively. Most will eventually recognize they need their job and like it enough to stay. From there, you build back to an effective team.

It is human nature to resist change. We become comfortable with the same old routines, the familiar things the way they are. Resistance to change is usually the result of fear: fear of the unknown, of being unable to learn to use new equipment or new technology, of reduced job security or status, of suffering some financial loss, of losing power or control, or fear of a change in workgroup relationships. Obviously, given today's workplace, many of these fears are justifiable (Figure 12 – 10).

Employees undergoing changes may not always be overt in their resistance. Instead, resistance may appear in the form of apathy, reduced morale, constant griping or turnover. So how do you deal with this?

Gaining Acceptance of Change

Instead of seeing change as a problem of overcoming resistance, Supervisors should learn to create an attitude of acceptance of change (Figure 12 – 11). This is a much more positive way of dealing with inevitable change. It is also realistic: if we're going to work together, we need to accept the reality and go from there. If we constantly resist the reality or talk nostalgically about the way things were, we cannot deal with the work that needs to be accomplished. Our method for overcoming resistance to change,

Figure 12 – 10. Fear Factors and Change

Things people fear:

- The unknown
- Being unable to operate new equipment
- New duties and responsibilities
- Being unable to learn new technology
- Layoff
- Unionization
- Deunionization
- Reduced job security
- Suffering some financial loss
- Takeovers
- Reduced job status
- Losing power or control
- A change in workgroup relationships
- New management
- New work location
- Different hours
- Shift changes
- Younger workers
- Older workers
- Affirmative action
- Being fired
- Firing

You can probably add a few:

then, is based on gaining acceptance of change. The following points will help you to achieve this acceptance.

- Familiarize yourself with the change that is to occur. If the change is coming from upper management, find out what is causing the change, the rationale behind it, the purpose and the benefits. PLAN! Try not to be caught by surprise.
- Anticipate the concerns your employees may have regarding the change. Just because you or upper management view the change as a good one doesn't necessarily mean your employees will initially embrace it with open arms. You may well have your own doubts. But, as a Supervisor, you are management. It's your job to explain and defend. To do this, you have to identify the gains and losses that go along with the change and anticipate employees' reactions.
- Communicate the change to your employees. Help develop an appreciation of the gains so the losses will be more tolerable. Explain the factors that led up to the change. Demonstrate the need for the change. Allow as much participation in the change as possible. This shows respect for employees, boosts their self-esteem and builds commitment. Allow for questions and discussion and listen actively. Clearly communicate the effective date of the change.
- Allow as much lead time as possible before implementing the full change. Provide any training necessary to carry out the change.
- Implement the change. Provide ongoing support, training and empathy. Monitor the change. Look for any unexpected problems or reactions to the change, and address them accordingly. Set dates and times for interim reviews.
- Evaluate the implementation of the change. Are the desired results being achieved?
- DON'T introduce change for change's sake.
- DO use your planning and decision-making skills throughout the change process.
- DON'T let the change manage you.
- DO manage change.

CONCLUSION

Today's business and service organizations are complex. With technology, competition, mergers and acquisitions, the environment seems constantly in flux. Therefore, Supervisors need skills and tools to cope with all these changes. Planning is a major tool, and the decision-making process is a skill that must be developed. The Supervisor of today and tomorrow must be prepared to:

- Plan
- Make decisions
- Solve problems
- Manage change

Figure 12 – 11. Gaining Acceptance of Change

- Familiarize yourself with the change that is to occur.
- Anticipate your employees' concerns.
- Communicate, explain and defend the change to your employees.
- Explain what led up to the change.
- Demonstrate the need for the change.
- Allow as much participation in the change as possible.
- Clearly communicate the effective date of the change.
- Allow as much lead time as possible before implementing the full change.
- Provide any training necessary to carry out the change.
- Implement the change.
- Evaluate the implementation of the change.

Chapter 13

Developing Employee Skills and Careers

OVERVIEW

Everything discussed thus far focuses on how to be a good Supervisor. Underlying all of that is the impact of good supervision on employees. In this chapter we focus on additional areas that will help you, the Supervisor, develop your employees to their fullest potential in the organization. The specific areas discussed include:

- Selection—matching employee to job
- Orientation—helping the new employee learn about the job, the company, the general rules, safety rules and procedures
- Training—implement on-the-job training and other learning experiences
- Coaching—assist and encourage individual employee development, and tips on managing stress, change and the introduction of new methods and procedures.

These areas are critical to the development of employees because they represent points where most people need help. Your job, as always, is to assist your employees. Well done, it is the extra boost that can help an employee go far in work-related personal development.

SELECTION

Chapter 9 discusses the selection of new personnel in great detail. Here, in Chapter 13, we are dealing with new hires, transfers, promotions and existing employees. It is worth repeating, however, that insuring a safe and productive workforce begins by selecting the appropriate employees in the first place. The selection process is obviously not foolproof, but using the techniques described in Chapter 9 certainly will increase your chances of hiring, promoting or accepting transfers of people who can become your best employees.

When selecting a new employee or trying to match employees to jobs, think of both long-range company plans and your immediate goal of finding the employee who is right for the job now. Thinking more long-term, you will want to hire employees who show potential for development within the organization. These are the people who can grow into a more responsible position. They are the people you can depend on not just to do their jobs, but to do them safely and be trusted assistants who can be promoted. In this context, select employees who:

- Have personal goals and characteristics that match the organization's, both immediately and long-term.
- Are motivated to do a good job and demonstrate interest in growing with the organization.
- Have a good safety record. Past performance is usually a good indicator of future performance. Someone with a poor safety record may need additional safety training and will require more time to grow. In some cases, such a person may have to be let go (Figure 13 – 1).

To determine whether employees have these characteristics, you have to know about your company and how to read the employee's record. You must know the short- and long-range company goals. Will the organization be implementing new technology that will require new skills? If so, you'd better look for them in those you hire now.

Ask applicants what their personal career goals are now and where they see themselves in the next 5 – 10 years. If they want to just *Do my job*, is that what you want and need? Ask the applicant if there are new skills they are interested in learning. See if you have a current and future match.

Look carefully at the employee's application or record. Examine work patterns. Does the person job-hop? Has the applicant done her/his homework? Does the employee know at least a little about your organization, such as what product or service it provides? If the applicant is currently working for the company, does he/she understand some of its structure? When checking employment references, always ask questions regarding the employee's safety record, training and involvement (Figure 13 – 2).

ORIENTATION

We urge Supervisors to orient new employees to make them feel welcome and valued members of the Supervisor's team. First impressions are usually lasting impressions. Orientation, done well, gives the employee a good first impression of the Supervisor and the workplace. Orientation reduces the employee's search for information, a search that can be frustrating and burdensome to the new employee. Orientation, expressing expectations, rules and procedures, can also help:

- Reduce absenteeism and turnover
- Prevent performance problems
- Set a pattern of good communication between the employee and Supervisor

Making safety indoctrination part of orientation is critical to effective workplace supervision. It gives the new employee essential information, signals that safe work procedures are a priority and helps reduce the occurrences of safety violations and accidents. For example, while showing the work area to the new

Figure 13 – 1. Some Standards for Selection

Select employees who:

- Have personal goals and characteristics matching your company's
- Are motivated to do a good job and show interest in growing
- Have a good safety record

Figure 13 – 2. Matching Potential Employee and Job

What are your company's short- and long-range goals?

Is there new technology requiring new skills?

What are the applicant's personal career goals?

Where does the applicant see himself/herself in the next 5 – 10 years?

Are there new skills the applicant is interested in learning?

What does the employee's application or employment history reveal?

What are the applicant's work patterns?

Does the person job-hop?

Do the references check out OK?

What is the applicant's safety record?

employee, the Supervisor should point out particular equipment or potential hazards that need special care and attention to safety.

Who's Involved in Orientation?

There are a lot of people who should be involved in the orientation process (Figure 13 – 3). First and foremost is the new employee. If possible, the new employee should be given an individualized orientation rather than as part of a group. Personalized orientation should occur right away. The Supervisor is the key person to orient the new employee. Although many others will often be involved in the process, the Supervisor should have the primary responsibility for the new employee and should orchestrate the whole process.

One of the first things to do is to assign a mentor (helper, buddy, experienced worker) for the new employee. The mentor should be selected from the Supervisor's own workgroup to assist in the orientation process. Choose someone who is experienced, has good technical skills and good communication skills. Being a mentor can be viewed as a reward, so it is best to choose a different mentor every time there is a new employee.

There is another group of people who should be involved: the actual workgroup that the employee is joining. The workgroup should be told in advance the new employee's name and starting date. Having your co-workers call you by name as soon as you arrive is quite a welcome. The workgroup should be reminded to help welcome and make the new employee as comfortable as possible. In spite of all the years in between the two experiences, starting school and starting a new job have a lot in common. It's an adventure, but you feel strange in a strange place. A little welcome goes a long way.

In many organizations, Human Resources (Personnel, Employee Relations) is where the new employee begins the first day of work. The employee may only sign the necessary forms or may be offered a formal orientation program. The Supervisor should know what is covered with the employee in this orientation to avoid needless repetition. Sometimes, however, reinforcement of issues like duties, safety and assignments is important.

The Orientation Process

The orientation process should begin before the employee's first day. When the employee is hired, the Supervisor should notify the workgroup of the pending arrival, prepare and supply the person's workstation and/or locker and have the necessary equipment or uniforms ready to be handed over. The Supervisor should also clear her/his calendar as much as possible the first few days of the employee's employment to allow for some individualized attention. It is helpful to prepare a checklist of orientation points, tasks and assignments.

The employee's first day. Invite the employee into your office or to a relatively private area where you are free to talk. As a tip, avoid sitting with a desk or table between you and the employee. Try moving your chair around to the front of the desk or to the end of the table so that you are sitting right next to the employee. This creates a sense of equality and fairness.

Figure 13 – 3. Who's Involved in Orientation?

New employee

Supervisor

Workgroup

Human Resources Department

Put the employee at ease with a friendly welcome. Discuss the employee's previous experience and how it fits into this organization, or discuss personal interests. Describe the orientation process and what will be occurring the next few days. Ask if there are any initial questions and take the time to answer them. Let the employee know that you are there to help and introduce others who will be able to help, such as the mentor you are assigning to assist this particular employee.

Information to be covered the first day should include (Figure 13 – 4):

1. The employee's specific tasks, responsibilities and performance standards
2. Company and department policies, including work rules
3. Safety regulations and information about the accident prevention program
4. Hours of work, including overtime, compensatory time or flex time, pay rates and pay days
5. Policies on sickdays, reporting illness and tardiness
6. Vacation and holiday policies
7. Break policies and customs
8. Protocol, including titles and proper forms of address
9. The telephone system and directory, if appropriate
10. Dress code or expectations
11. Uniforms or personal protective equipment, such as shoes, hard hats, eye protection, special protective clothing
12. Reasons why protective equipment must be worn and penalties for violation
13. Probationary policy as applied to new hires, promotions transfers

Many companies require that the employee sign-off on receiving parts or all of the above information.

Give the employee a tour of the facility. Include such things as lockers, washrooms, eyewashes, safety showers, coat closet, cafeteria, break room, entrances and exits, message boards, bulletin boards, mail slots and the work unit location. During the tour, the employee should be introduced to co-workers. Do not give the employee any background about the members of the workgroup. Let the new employee form independent opinions of the co-workers. The Supervisor or assigned mentor should have lunch with the employee the first few days.

Take the employee to his/her workstation. Begin training the employee in the proper and safe operation of any equipment, then observe as he/she operates the equipment. Also familiarize him/her with any basic equipment or supplies, like the copy machine and where to find a pad of paper and a pen. Review and walk through the established emergency procedures, such as evacuation, fire, bomb threat.

Let the employee know about any scheduled meetings or programs. Explain whether the employee will be a participant or observer at these meetings. Describe any applicable formal training programs.

At the conclusion of the employee's first day, the Supervisor

Figure 13 – 4. The First Day: There's a Lot to Tell

The following information should be covered on the new employee's first day:

JOB

Specific tasks
Responsibilities
Performance standards

RULES

Company policies
Department policies
Work rules
Safety regulations
Probationary policy

HOURS

Hours of work
Days off
Overtime
Compensatory time
Flex time

PAY

Pay rates
Pay days
Deductions

BENEFITS

Medical
Life
Disability
Pension
Savings

ABSENCES

Sick policy
Tardiness policy
Reporting sick days
Vacations
Holidays
Breaks
Meals

CHAIN OF COMMAND

Protocol
Who's who
Titles

EQUIPMENT

Dress code
Uniforms
Personal protective equipment

should spend time with the employee to answer any questions. Again, express your welcome and tell the employee how glad you are to have him/her on board and express your confidence in her/him.

The first few weeks and months. Many of the things covered the first day will have to be reiterated, probably in more detail throughout the first few weeks and months. Remember how overwhelming your first few days on the job were. Try to provide ongoing training specific to the employee's job responsibilities. Allow for a period of adjustment and allow for mistakes to occur without fear of retribution or punishment. Provide periodic positive feedback sessions to catch any problems early.

Give the employee more detailed information about the organization, its structure and goals and explain how the employee's responsibilities fit into the big picture. The first six months of a new job are critical so don't assume the employee will pick up everything the first few days or weeks. The learning process takes time.

Safety Training

An important component of any orientation program is safety indoctrination. Teaching safety rules and procedures early in the new employee's career emphasizes that the company views safety as important. However, giving the employee the list of safety rules isn't safety training. Safety instruction should follow the same training steps used to teach job skills (see the next section, Training, in this chapter; see also Chapter 9, Safety Training, in *Accident Prevention Manual for Industrial Operations: Administration and Programs*, National Safety Council, 1988.

The importance of safety training can also be illustrated by these statistics from *Accident Facts*, National Safety Council, 1988:

- While you make a 10-minute safety speech, 2 persons will be killed and about 170 will suffer a disabling injury.
- On the average, there are 11 accidental deaths and 1,000 disabling injuries every hour during the year.
- A work accident resulting in death occurs every 47 minutes and a work accident resulting in injury occurs every 18 seconds.
- Work accidents cost 42.4 billion dollars in 1987, including lost wages, medical expense, insurance administration, fire loss and indirect loss.
- In 1987 time lost due to work injuries was 35,000,000 days.

These facts leave no doubt about the importance of safety in the workplace, and the need for early safety training of the new employee.

Four steps are involved in safety indoctrination (Figure 13 – 5):

1. Prepare the employee for a safe work attitude.
2. Explain and personalize a safe work attitude.

Figure 13 – 5. Four Key Steps of Safety Indoctrination

1. Lay a foundation for safe work attitudes.
2. Help personalize safe work attitudes.
3. Adopt safe work attitudes and habits.
4. Reinforce safe attitudes and habits.

3. Provide opportunities to practice a safe work attitude and habits.
4. Follow up by reinforcing the safe attitude and habits.

Preparing the employee. Laying the foundation for a safe work attitude involves orienting the employee to the workplace safety rules and the expectations of the Supervisor and company. Company procedures may require that the meeting date and topics discussed be documented and signed by the Supervisor and employee. The Supervisor should schedule a meeting with the new employee the first day on the job to discuss safety. The Supervisor should explain to the employee how important safety is in the workplace. The rules should be explained thoroughly. Any safety equipment should be explained, and the employee should be trained how to use it. Any potential chemical hazards in the workplace must be shown to the employee, as well as the dangers and methods of protection. (This is to comply with the Hazard Communication Standard of OSHA, commonly called the right-to-know standard.) The employee should be given the opportunity and be encouraged to ask questions before being permitted to work with the equipment or perform the procedure.

Explain and personalize safety. The Supervisor should show the employee both the benefits of following the safety rules and the negative consequences of not following them. Adopting safe work habits should be included in performance expectations and thus reflected in performance appraisals and merit increases. Point out clearly that failing to adopt safe work practices can result in disciplinary actions, disabling injuries or accidental death. Requirements should be acknowledged in writing by the employee. The Supervisor should share both the positive outcome of cases in which employees have successfully adopted safety habits and the negative consequences of cases where safety practices weren't followed. The Supervisor can also give a demonstration of the use/application of all safety procedures, equipment and safeguards. Check off each item or procedure completed and give the date. Both the employee and the Supervisor should sign the checklist. Periodic follow-up testing or review should be done for critical safety areas.

Provide opportunities to practice. The employee should then be allowed to practice safety techniques. Work with the employee to see that equipment is used properly and safely. See that the employee is properly using the personal safety gear. Check frequently to see that the employee is keeping the workspace clean and free of any debris or potential hazards. Reinforce and praise the safety habits and correct anything you see going wrong. In this way, you aid the employee in making a safe work attitude part of her/his work personality.

Follow up. Continue to reinforce the employee's safe work habits beyond just the orientation phase. Meet with employees from time to time to discuss safety. Give workgroup training programs to inform them of any changes in the workplace or to train them on new equipment. Include safety performance in the appraisal process. Give positive feedback to employees who demonstrate good safety habits and take prompt, corrective action with any employee who demonstrates unsafe practices.

TRAINING

Few people can step into a new job or apply a skill they have never used before and perform successfully the first time. Giving or receiving training is an ongoing part of any job. As a Supervisor, you are responsible for making sure that your employees are properly trained for the work they are doing or are about to do. This may involve your conducting actual training of your employees, or it may be a matter of making sure that your employees attend the appropriate training programs. Proper training will help develop your employees by teaching them new skills that will help them advance their careers. Supervisors also promote training by setting a good example.

On-the-Job-Training (OJT)

There are four key steps involved in effective on-the-job training, much like those involved in effective safety indoctrination (Figure 13 – 6).

Prepare the employee for OJT. Explain the purpose of the training and why it is important. If the training involves new procedures that are being introduced in the workplace, explain why the company has adopted these new procedures. Express your confidence in the employee's ability to learn and master the new skill. Allow the employee to express any reservations or questions he/she may have. Explain what will occur during training.

Explain the process. Explain the steps in the process of the new skill and then demonstrate how it is to be done. Provide a positive model rather than showing the employees how not to do it. Use as many techniques as possible to show the employee how to do it: hands-on demonstration, video tapes, pictures, diagrams. Remember, seeing is believing. Remember also that you are constantly providing OJT by the example you set every day, so be sure you are a positive example. Explain the quality standards that are expected once the employee has completed the training process. Break things down into simple steps, so that the employee can have success along the way and build on success (Figure 13 – 7 and Chapter 3).

Don't expect anyone to master a skill by simply reading an instruction manual. Written directions provide background information that is important, but learning occurs in application of a skill. Just as reading this book alone won't make you a good Supervisor until you apply and practice what is covered, learning a skill cannot be completed just by reading about it. Applying and practicing the skills and techniques presented will help make you a good Supervisor. Applying and practicing any skill is the only path to mastery.

In an environment where technology is constantly changing, OJT is a must. Training is an investment in productivity. If you don't teach new skills, techniques or procedures, your people will not have the skills necessary to improve production and productivity. Lack of or improper training will negatively impact performance and job satisfaction. Many accidents occur because someone wasn't trained properly.

Figure 13 – 6. Four Key Steps for OJT

1. Prepare the employee.
2. Explain the process.
3. Provide for tryouts.
4. Apply the new skills gradually.

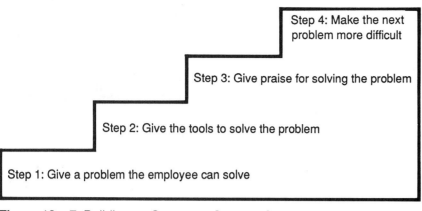

Figure 13 – 7. Building on Success—Steps to Sharing Your Work Load

Provide for tryouts. Allow the employee to practice the new skill. As always, reinforce the positive and correct anything that goes wrong. Don't punish or criticize the employee while she/he is trying to learn. Let the employee practice with your assistance. A little practice may not make perfect, but practice, practice and more practice usually does.

Follow up. Gradually, have the trainee apply the new skill alone without your assistance. Continue to observe and provide support and reinforcement.

Training Programs for Employees

Don't send someone to training just because the training program exists. If one of your employees is generally doing a good job but needs development in a particular area, consider a training program. For instance, your secretary may have good clerical skills but sounds gruff and discourteous when he answers the phone. If you have customers getting turned off by your secretary's phone manner, you may wish to send him for a Phone Skills Workshop. Another example may be if your secretary is doing an excellent job at all of her current tasks, but has indicated an interest in learning a new computer program that would help her do the job even more efficiently and give her the skills necessary for a future promotion. Or, perhaps one of your self-starter employees seems capable of moving on to more responsibility but simply lacks some skills and that specific training is not available at the company. Find a place where the training is available and send her to it. Equal employment opportunity includes training as it relates to promotion or advancement.

Formal training programs. Formal training programs, done by someone other than the Supervisor or mentor, are also important to the development of your employees. Your company may have a Training and Development Department that offers programs on-site. Additional programs are offered off-site in the community, often by local community colleges and high schools. These programs can be used to help your employees master their current job or help them develop skills that can be applied in a new job or a promotion.

Find appropriate training. Find the appropriate training to fit the need. In-house training is usually the most cost-effective, so check there first. When deciding on the appropriateness of any training program, consider these factors (Figure 13 – 8):

- What is the stated objective of the program?
- Does the program specify what skills will be taught?
- Who are the instructors? Are they experienced? Do they have the proper background or certification necessary to teach the course?
- What company or organization is sponsoring the training? Is it reputable?
- Who are the references or clients listed? Can you call them? Can you talk with others who have attended the training program?
- What training methods are used? Is there a combination of lecture, demonstration and skill practice?
- Where is the training located? Is it convenient and cost-effective?
- What is the length of the program?
- What type of feedback will be provided to the Supervisor?
- What is the cost compared to the benefit to be gained?
- How will you measure the effectiveness of the training?

Discuss training needs with others. Your company training specialist should be able to help you locate programs to fit your needs and answer some of the questions listed above. Determine if there is tuition assistance offered by the company. If you have no training specialists, or they are far removed from your location, try networking the information through Supervisors at the facility or from neighboring companies.

Discuss goals and objectives. If you send your employees to training, take time before they attend to discuss the goals of the training and what they can expect while in training. Tell them to be prepared to report on the program and what they learned from it upon their return. When they return from training, sit down with the employees and discuss what they have learned. Offer your assistance in helping them apply their new-found knowledge. Use praise to reinforce the employees' demonstration of the new skills. Encourage the employees to share their new knowledge and skills with others who were not able to attend the seminar.

COACHING

What is coaching? Coaching in the workplace may be a little different from the athletic coach who comes to mind, although there are some similarities. In the workplace, coaching is helping employees in a positive manner and developing them to full potential. Positive is the key here. Coaching does not involve yelling at your players every time they do something wrong or repeatedly playing the film of the play that went sour. Instead, coaching involves cheering on your employees and providing supportive behavior.

Key elements of coaching. The Supervisor must perform

Figure 13 – 8. Finding Appropriate Training

Always check the following factors:

- Program objective
- Skills taught
- Instructor's experience
- Sponsor's reputation
- References, client lists, word of mouth
- Training methods
- Skills practice time
- Location of training
- Length of the program
- Feedback provided
- Price compared to benefit

the following key elements of coaching in the workplace (Figure 13 – 9):

- Establish goals and performance standards or requirements.
- Provide nonthreatening advice and support.
- Listen empathetically to employees and probe for how they feel about their work and themselves.
- Carry out coaching in a nondistracting environment, if possible.
- Work to inspire trust, confidence and respect. One of the key elements of inspiring trust, confidence and respect is extending it to those you supervise.
- Try hard to be in a helping role rather than the boss with a critical eye. Any criticisms offered should be to help, improve or prevent, not to punish. They should be accompanied by advice on how the employee can build on existing skills and knowledge to do the job right.
- Use the good communication skills detailed in Chapter 4.
- Be patient. A good coach is patient, very patient even when it is the most difficult time to be patient.
- Use coaching to reinforce new skills, improve performance problems, or solicit ideas to solve a departmental or organizational problem.

Coaching and career development. The purpose of developing your employees' careers is to enhance their own job satisfaction which, in turn, benefits the employer. If an employee begins to feel trapped or dissatisfied with the current job, productivity, safety and quality might decrease or the employee might leave. No one wants such decreases and the loss of a valued employee can be costly. As a Supervisor, you should always be on the look-out for ways to enhance employees' current jobs or develop them to advance within the organization. This is different from performance counseling as described in Chapter 10. Development involves helping employees who are performing quite well but demonstrate interest or potential in growing further (Figure 13 – 10).

As a Supervisor, you can take the following steps to develop your employees (Figure 13 – 11):

1. Get to know your employees. Find out how satisfied they are with their current job and what their job goals are. Sudden performance problems from a high performer may mean he/she is just bored and needs a new challenge.
2. Know your organization's policies and procedures for career development. What are the rules concerning internal transfers or promotions? What training opportunities are available? What type of educational assistance is available? Is there career counseling available in the Human Resources Department? Find out what skills the organization will need one, three, five, 10 years down the line. Are there current skills being used that will be considered obsolete in a few years?
3. Meet with your employees at least annually to discuss

Figure 13 – 9. Keys to Good Coaching

The workplace coach must:

- Provide nonthreatening advice and support
- Listen empathetically
- Coach in a nondistracting place
- Inspire trust, confidence and respect
- Extend trust, confidence and respect
- Stay in a helping role
- Don't be the critical boss
- Use good communication skills
- Be very, very patient
- Use coaching to reinforce new skills, improve performance problems or solicit ideas to solve problems

Figure 13 – 10. Employee Development

EMPLOYEE DEVELOPMENT involves helping employees who are performing quite well but demonstrate interest or potential in growing further.

development plans. This should be separate from your performance appraisal goal-setting meetings. Outline goals and objectives. Communicate what you discovered in exploring your organization's career development policies. Outline the concept of career development planning and action steps. For instance, your secretary may have expressed interest in moving up to an administrative secretarial position. In order to be qualified, however, he would need some experience with the company accounting and budgeting procedure. What can be done?

4. Develop action steps. Using the above example, the action steps that the two of you agree on could be:
   - The secretary will spend two hours a week with the department's administrative secretary learning how to process invoices, vouchers and purchase orders according to company (or departmental) procedure.
   - Initially, practice standards should be set. These should then be replaced with the performance standards in place for the administrative secretary position.
   - Once familiar with the process, the secretary would perform this task in the administrative secretary's absence (vacation, illness, etc.).
   - The secretary will take an in-house training program to learn the software package used for the organization's budget tracking system.
   - The secretary will apply for educational assistance to enroll in the local college for a basic accounting course.

5. Coach and encourage. Once the development plan has been outlined, it is the Supervisor's role to coach and encourage the employee to carry out the plan. Feedback and interim review of the action plan is important.

Development can be used on a much lesser scale as well. For instance, offering an employee the opportunity to be a mentor for a new hire can be enough to break up the routine and express confidence in an employee who has been doing a good job. Assignment of an employee to a new project can have a similar effect. Ultimately, of course, it is the individual's responsibility to develop his/her own career, but the Supervisor can certainly lay the groundwork and be a supportive coach along the way.

MANAGING STRESS

First of all, just what is stress? Webster's definition (New Collegiate) is: *Constraining, urging, or impelling force, constraining power or influence; pressure; urgency. Psychological stress is a set of interactions between a person and the environment that result in an unpleasant and emotional state, such as anxiety, tension, guilt or shame.* Stress in the workplace is one of the most common problems faced today. Dozens of articles have been written on the subject. Executives feel stress, and many articles focus on executive stress. But stress reaches throughout the workplace, from executives to managers to Supervisors to employees.

All workplace stress is not bad. For example, you may feel

Figure 13 – 11. Supervisor's Tasks in Developing Employees

Get to know your employees.

Know your organization's career development policies and procedures.

Meet with your employees at least annually to discuss development plans.

Develop action steps.

Coach and encourage.

stress when handling hazardous materials, but this very stress can be a positive pressure making you aware of the necessity to carefully follow safety procedures. Similarly, many people seem to perform best when under the stress of a deadline or production schedule. So, stress can be bad, or stress can be good. The important thing, however, is to try to manage stress.

This section focuses on employee stress and how you, the Supervisor, can help manage it. (Of course, you can apply many of the identified problems and suggested techniques for managing stress to yourself! See Chapter 14 on stress management for Supervisors.)

Employee stress factors. Employee stress factors in the workplace include (Figure 13 – 12):

- Safety hazards, such as chemicals, dangerous machinery, cluttered or dirty workstations, noise, threat to life or limb. If employees have to stumble around an unsafe workplace, they're under stress.
- Nonsupportive or unappreciative Supervisors. Yes, your behavior can cause stress in your employees. Remember the fact that you are a role model for those you supervise. If you work unsafely, so will your employees. If you're stressed-out, they will feel it and respond in kind. If you harass, they'll feel that as stress.
- Routine or boring tasks. Parts of every job are routine and boring. The job of an airline pilot sounds exciting. But imagine what it takes to stay alert flying a jumbo jet when the computer is actually in control and all you're doing is sitting and watching. Boredom causes stress.
- Job dissatisfaction. If an employee is unhappy with his/her job, that unhappiness weighs on the employee who will feel stressed.
- Low pay. If an employee cannot meet her/his basic needs because of low pay, that will carry over to the workplace in the form of worry and resentment, and that produces stress.
- Overload. When numerous demands are placed on an employee all at once, the employee will feel stressed. While some people perform well under high stress conditions, no one can survive in a state of constant, unrelenting stress.
- Little or no control over the work situation. If the employee doesn't know what's coming next, can't control his/her own pace, can't determine the best way to solve a problem and is always at the mercy of someone else's demands, the employee will feel stressed.
- Lack of pertinent or necessary information, equipment or training to get the job done. If an employee has been assigned a task and doesn't know how to do it, or doesn't have the right tools, the frustration and concern for failure will produce stress.
- Negative environmental factors. These can include many things often overlooked, such as poor lighting, badly designed workstations, desks positioned where people constantly interrupt work, labs where workers must move from one area to another, lack of heating and air condition-

Figure 13 – 12. Employee Workplace Stress Factors

- Safety hazards
- Nonsupportive Supervisors
- Unappreciative Supervisors
- Routine or boring tasks
- Job dissatisfaction
- Low pay
- Numerous simultaneous demands
- Everything's a crisis environment
- Little or no control over work situation
- Lack of necessary information
- Negative environmental factors

ing, improper regulation of heating and air conditioning temperature. For example, at a food processing production facility the company provided piped-in FM music over the intercom system to try to relieve boredom. The employees all agreed on having music, but there have been continuing major disputes among employees over what kind of music: rock, hard rock, country, oldtimers, and so on! Can this produce stress? Absolutely.

Off-the-job stress. Many employees bring off-the-job stress to the job. For example, if an employee is going through a divorce, a death in the family or problems with a child, he/she will be under stress and will inevitably bring some of that stress to the job. However, off-the-job stress is best treated as a performance problem (see Chapter 10), or, depending on its interference with work, a discipline problem (see Chapter 11), or something that will need to be referred to the company's Employee Assistance Program (again, Chapter 11). The causes of off-the-job stress are completely beyond your control. You are faced with the symptoms when they interfere with performance on the job, but there is little or nothing you can do to relieve the causes.

Impact of stress on safety and productivity. The impact of stress on safety and productivity is best illustrated by National Safety Council studies. These indicate, for example, that in 1986 stress was directly or indirectly responsible for 60 – 80 percent of industrial accidents. Statistics also show that stress-related workers' compensation claims are increasing at a rapid rate. In 1989, 26 states allowed for compensatory claims for employees who suffered from mental trauma due to job-related stress. ("Health Action Managers," February 10, 1989, Kelly Communications, Charlottesville.)

National Safety Council studies have shown that the effects of stress can lead to costly mistakes, decrease in productivity, absenteeism and turnover. That means undesirable stress in the workplace is a costly matter and one that needs to be addressed.

Identifying stressful behavior. Supervisors must be aware of the warning signals of stress and learn to deal with them in ways that will help to relieve some of the stressful behaviors. Employees often show the following symptoms when they are experiencing stress:

- Inattention to detail
- Easily distracted
- Easily frustrated
- Outwardly frustrated
- Unnecessary risk-taking
- Fatigue
- Agitated appearance
- Increased tardiness
- Increased absenteeism
- Apathetic attitude
- Argumentative attitude

The symptoms described are all observable. They also imply a change in behavior. If Joe always has a negative attitude, it is probably not stress-related, but it represents a behavior or performance problem and should be treated as such. If, however, Joe normally was cooperative and suddenly begins exhibiting negative attitudes or behaviors, it can be stress-related.

Note that all of these stress indicators could be symptoms of stress related to work or to factors outside work. The first step, then, is to identify the source of the stress by talking with the employee in a coaching manner. If the cause of stress is not work-related, use the performance and corrective counseling approaches mentioned earlier or refer the employee to the EAP. If the cause is work related, read the next section.

Helping to relieve stressful behavior. The idea, of course, is that Supervisors can help alleviate employee stress. Can this really be done? Yes, with effective stress management. Note, however, that the active word is alleviate, not eliminate. The goal is to reduce stress, not to eliminate it. The Supervisor probably cannot do the latter and still maintain a productive workplace!

Here is a list of the main things you can do to alleviate stress in the workplace (Figure 13 – 13):

- Don't be a stress-carrier to your employees. If you routinely act like the place is on fire, you're communicating constant stress. It does pay to do this when the place is on fire, but that should be it.
- Offer your employees variety in job tasks. If tasks are routine and repetitive, give periodic breaks. Even a simple switch in workstation or location will help. Or find out if your organization allows job rotation.
- Survey the physical environment. Eliminate safety hazards as much as possible. To the extent possible, be sure the workplace is clean, well-lit and free from undue distractions, including excessive noise, heat, humidity and air pollution.
- Emphasize safety rules and compliance.
- Encourage teamwork, cooperation and participation.
- Plan work activities so that rush jobs are the exception rather than the rule.
- Apply the Supervisory skills discussed throughout this book: be a good communicator (i.e., listener), treat employees fairly and equitably, provide performance expectations and ongoing feedback, be sure your employees have the resources necessary to get the job done, offer development opportunities and keep a safe workplace.

Encouraging others to adopt stress-management techniques. Stress is inevitable, and everyone must learn to deal with it. Before you can do much about encouraging others to adopt stress management techniques, you must, of course, adopt your own stress management techniques (discussed in Chapter 14) and set a good example. Given that very important first step, here are some of the things you can do to encourage effective stress management among your employees.

- Allow, encourage and work with your employees to participate in or start any company-sponsored wellness ac-

Figure 13 – 13. How to Help Relieve Stress

- Don't be a stress-carrier.
- Offer variety in job tasks.
- For repetitive tasks, offer periodic breaks.
- Regularly survey the physical environment.
- Eliminate safety hazards.
- Try to provide a clean, well-lit workplace.
- Emphasize safety rules and compliance.
- Encourage teamwork and participation.
- Plan work activities.
- Make rush jobs the exception rather than the rule.
- Apply the supervisory skills discussed in this book.

tivities, such as fitness programs, nutrition programs, smoking cessation programs or stress management classes.

- Communicate to your employees the availability of the company's Employee Assistance Program, if one exists, or any alternative programs available.
- Encourage employee participation in community-based wellness activities sponsored by community colleges, high schools, churches and synagogues, and so on.

The following case study illustrates a situation of extreme tension in which the Supervisor has to defuse a tense situation between an employee and an "owner/customer."

CASE STUDY—TOUGH DAY AT THE EXCHANGE

The people involved in this case study are Garrison Carson, an Exchange employee; Wendy Wilson, the Stock Pit Supervisor responsible for supervising Garrison's work; and Langhorne Barnes, a Stock Trader and a member of the Exchange Governing Board.

Garrison Carson has worked on the Stock Exchange floor for six months. His job is to listen carefully to the stock deals called out by the traders in the pit (the small and very crowded areas where the dealers actually make a deal to buy or sell at a price they agree to). He then has to call in the trade price by walkie talkie or hand signals to other Exchange employees sitting in galleries above the pit. They, in turn, enter the data into the computer, and the deal flashes over the electronic board, much like sports scoreboards. The big difference in these scoreboards is that at the Exchange, the numbers change constantly. A mistake in recording the correct number can cost thousands of dollars to the traders. In spite of that fact, Exchange employees are only allowed to listen; they cannot interrupt traders with questions because that would slow down the trading. A normally tense set of circumstances!

Monday was a particularly tense day. Stocks were falling, and the trading was very brisk. Garrison was aware that Langhorne Barnes, a very wealthy trader and a member of the Exchange Governing Board, was quite agitated this day. He listened carefully to the terms of the deal Barnes was making and called in the numbers. He was listening to another deal by different traders when he felt a hand on his shoulder and was whipped around to find himself facing Langhorne Barnes. The following conversation took place. It reflects the heat of the Exchange and the turbulent personalities often found there.

BARNES: *You damned idiot! You reported that number wrong, and I'm losing thousands of dollars. Can't you hear? Were did you flunk out of school? Get the damned stuffing out of your lousy ears and get the hell out of here!*

During all this, Barnes was still holding onto Garrison and had begun shaking him.

GARRISON: *I'm sorry, sir, but I reported that trade correctly. That's what I heard, and that's what I called in. Please take your hand off me now.*

Garrison tried to shrug off Barnes' hand. Supervisor Wendy Wilson noticed the commotion and literally ran over to the pit area, where she stepped between the two men.

WENDY: *Garrison, please stand back. Mr. Barnes, what's the problem? Please let go of Mr. Carson and tell me what's going on.*

Supervisor Wilson firmly removed Barnes' hand from Garrison and also give Garrison a gentle push away.

BARNES: *That damned moron can't even get numbers right. I lost five thousand dollars over his lousy mistake. Get rid of him. I want him fired right now.*

WENDY: *Mr. Barnes, there must be some mistake. Garrison has been doing a good job here. Let's see if . . .*

BARNES: *Don't try and excuse his incompetence to me, young lady. I want him out of here, or I'll demand a stop in trading. Now! Or you're next!*

GARRISON: *Wendy, I'm trying to be polite, but this guy is just . . .*

WENDY: *Garrison, please leave the trading floor and wait for me in the locker room. Mr. Barnes, I will look into this and get all the facts, then I'll get back to you.*

At this point, Barnes grumbled and went back to his work. Supervisor Wilson stayed in that pit area herself, since there was no one else to replace Garrison. That meant she couldn't meet with Garrison right away, and he had to feel pretty abused and abandoned. Yet, she had stopped the fight, separated the parties and allowed the work to go on. Tense situations such as this occur frequently in the various Stock and Money Exchanges. Given the tension and almost impossible situation, what did Wendy Wilson accomplish? What didn't she accomplish?

CONCLUSION

Employee development involves many steps: selecting and matching the right employee to the right job, proceeding through the first very important days and weeks when the new employee comes to the job, orienting the new employee, providing safety and job training. Workplace stress is inevitable but you, the Supervisor, can help relieve some of that stress through methods listed in this chapter. Throughout this chapter, it is apparent that much of what is discussed relates to various earlier topics, such as safety, communication, planning, performance and discipline. The best way to develop employees is, of course, to be a good Supervisor.

Chapter 14

Developing Your Career

OVERVIEW

Chapter 13 explored how you can help develop the careers of your employees. Now it's time to focus on career development and the Supervisor. Few people progress in their careers successfully without planning and taking action to get where they want to go. Life rarely presents prepackaged golden opportunities. If you don't plan and take the time to develop your career, no one else will do it for you, and you may find yourself at mid-career not being where you really want to be. This chapter focuses on how to take control of your career. It gives you some tools to use to help develop your career. The following areas are covered:

- Assessing your priorities and goals
- Designing a development plan/strategy
- Influencing your boss
- Time management
- Wellness

Taken together, these five areas will provide you with the tools necessary to develop your career and take significant control of your worklife.

ASSESSING YOUR PRIORITIES AND GOALS

Developing your career plans is similar to the work planning process described in Chapter 12. Without an objective or identified priorities, plans go astray. The first step in your career development, then, is to establish your personal priorities and goals.

Self-Assessment Tools

Several self-assessment tools are available to help you determine your priorities and goals. The popular book, *What Color is Your Parachute?* is published annually and contains several useful exercises to help you determine priorities and goals. Your company's Human Resources Department may also have some books or skills inventory tests to get you started. Find out what other career planning services your company offers. Remember, it is in the interest of your company to have you increase your skills and capabilities to perform more responsible jobs in the company. They have a vested interest in helping you if your effort is focused on staying with the company. Your local library is also a good source for career development information.

Determining Goals

Several factors determine what our goals and priorities are. Here are just a few to consider:

- What do I particularly enjoy about my current job or jobs I have held in the past?
- What do I especially dislike about my current or past job? Figure 14 – 1 gives you a checklist of likes and dislikes to help you do a self-assessment.
- What do I especially enjoy or not enjoy about my current or past boss? Leadership style? Method of treating employees? Ability to delegate?
- What are my strong points or highly skilled areas? Initiative? Willingness to take on whatever is assigned?
- What are my weaknesses? Procrastination? Blame others? Worry a lot? Unable to delegate?
- What skills am I interested in developing? Communication? Confronting discipline problems? Delegating?

Figure 14 – 1. What I Like and Dislike about My Job

I LIKE	I DISLIKE	
( )	( )	SUPERVISORY RESPONSIBILITY
( )	( )	RELATIONSHIPS WITH MY CO-WORKERS
( )	( )	RELATIONSHIPS WITH MY SUBORDINATES
( )	( )	RELATIONSHIPS WITH MY BOSS
( )	( )	GEOGRAPHIC LOCATION
( )	( )	SMALL COMPANY
( )	( )	LARGE COMPANY
( )	( )	BENEFITS
( )	( )	PAY
( )	( )	OPPORTUNITIES FOR ADVANCEMENT
( )	( )	AUTONOMY
( )	( )	FLEXIBLE HOURS
( )	( )	DAILY CHALLENGES
( )	( )	CONSTANT CHANGE
( )	( )	THE ROUTINE
( )	( )	SAFETY ENVIRONMENT
( )	( )	PHYSICAL SURROUNDING
( )	( )	INDOOR WORK
( )	( )	OUTDOOR WORK
( )	( )	ONGOING EDUCATION
( )	( )	TRAINING OPPORTUNITIES
( )	( )	WORKING WITH PEOPLE
( )	( )	WORKING WITH THINGS
( )	( )	UNION ENVIRONMENT
( )	( )	NONUNION ENVIRONMENT
( )	( )	COMPANY ATTITUDE ON ADVANCEMENT
( )	( )	OTHER

- What other positions or careers interest me? Manager? Technician? Sales?
- What are my family responsibilities and priorities? Children's or spouse's education? Summer cottage?
- Am I interested in continuing my education? High school? College? Graduate school? Technical courses?
- What are my short-term goals? For the job? For myself? Family?
- What are my long-term goals? For the job? For myself? For my family?

Figure 14 – 2 gives you a worksheet to develop the answers to these questions. Once you have identified the elements that are important and you are good at, rank them. Also list items that you really dislike or couldn't tolerate in a job.

Figure 14 – 2. How Do I See Things?

1. What do I especially not enjoy in my boss? List at least two things, even if you have a very good boss.

_____
_____
_____

2. What are my strong traits or highly skilled areas? List two traits and two skills.

_____
_____
_____

3. What are my weaknesses? List five.

_____
_____
_____

4. What skills am I interested in developing? List four or more.

_____
_____
_____

5. What other positions/careers interest me? List at least one.

_____
_____
_____

6. What are my family responsibilities and priorities?

_____
_____
_____

7. Am I interested in continuing my education? Where? When? What?

_____
_____
_____

8. What are my short-term goals? List at least three.

_____
_____
_____

9. What are my long-term goals? List at least two.

_____
_____
_____

aware of how people react to your new skills. Are you getting the response you want? Practice some more!

Taking Risks

Career development involves taking risks. Unless we want to maintain the status quo the rest of our lives, we will have to change from time to time, and changing means taking a risk. Risk-taking is scary for most of us, but it is a necessary, critical step to career advancement. You can minimize the amount of risk you must take and select the best risk options if you consider and plan carefully. These planning steps can help:

- Think about risks you have taken that have been successful. What made them successful?
- Think of some of the risks you took that didn't work out. Why did they fail?
- Observe people who are successful risk-takers. What do they do that you aren't doing?
- Determine the ways you avoid taking risks. Be honest with yourself and ask someone you trust to help you look at this issue. Why were you avoiding the change? How many times did your caution prevent your possibilities for success?
- Take calculated risks, not just risk for risk's sake. Be cautious, but understand that risk-taking is a critical and necessary step to advancement.
- Evaluate and monitor your risk-taking.

Taking Responsibility

Many people simply wait for opportunity to knock. It seldom happens. If you really want to advance, you must take responsibility for your development plan and how you will implement it. Chances are that no one will develop your career for you. Many people can help along the way, and many will if you demonstrate initiative. But ultimately, the responsibility is yours. Take the initiative to ask your boss to help you with your development plan. Your boss can provide valuable insight into areas that you can develop in your current position as well as future plans. Don't procrastinate about developing your career plan. If you are dissatisfied with your current position or feel the need to enhance your career, stop talking and complaining and start doing!

INFLUENCING YOUR BOSS

Like it or not, we all need our bosses. If you plan to stay with your current company, you won't get far without a good relationship with your boss. Even if you plan to move on to another organization, it will be much easier with a positive reference from your current boss. Don't assume the communication between you and your boss is a 50 – 50 proposition. Be willing to take all of the responsibility for keeping the lines of communication open and positive.

Assess Your Relationship with Your Boss

If you are going to be able to influence your boss in order to enhance your career development, you have to first take a look at exactly how the relationship between you operates.

- Do you interact with your boss the same way you interact with your staff and colleagues? Why not? Are you sure he/she wants it that way?
- Do you understand your boss as a person? Do you understand his/her job, demands and pressures? Take some time to think about that.
- What are your boss' pet peeves, prejudices, blind spots? Do you avoid stirring these, or do you sometimes goad a bit just to see the reaction?
- What is important to your boss? What are his/her priorities? For the company? For advancement? For your fellow workers? For you?
- Is there open communication between you and your boss? Think about how you would describe it to a neutral person.
- What nonverbal signals do you receive from your boss? Hasn't got time for you? Always busy? Ready to talk?
- Do you know if your boss is at his/her best in the morning, the afternoon?
- Do you avoid interacting with your boss when she/he is tired, angry, frustrated, preoccupied?
- Do you keep your boss informed on matters he/she wants to be kept informed of?
- Do you know what your boss's boss is like?
- Do you follow the chain of command?
- To what extent does your boss depend on you?

If you begin to build answers to these questions, you will have a profile of your boss and you. You should also begin to identify some areas where you need to change your behavior in order to have a better relationship. If you use the knowledge this simple checklist can give you, you will have the basis for improving the relationship and being able to influence your boss.

Tips on Getting Along with Your Boss

Remember that your effort is to influence, not to change. You will wear yourself out trying to change your boss, and in the end you won't succeed anyway. You can change yourself, however. The point is to have your boss as an ally who will be supportive of your career development efforts (Figure 14 – 4).

The boss' value. The first tip is to focus on the boss' value to the company, himself or herself, and you.

Looking good is important. Always work hard to make your boss look good. Every time you can help that happen you'll be helping yourself too. Show respect for your boss' management style. Understand that people follow a style because they see it as part of their self. Major efforts to change that style are threatening, and why should they put up with being threatened by someone below them in the hierarchy?

Try working on your active listening skills. The boss likes to

Figure 14 – 4. Very Condensed Tips on Getting Along with Your Boss

The Boss' Value
- Keep the boss looking good.
- Show respect for your boss' management style.
- Use active listening.
- A lot of enthusiasm goes a long way.

Taking Credit
- Always be accountable for your own actions.
- Share the credit for work well done.
- Work hard at being a team player.

Your Accessibility
- Be available.
- Demonstrate your loyalty.
- Learn good timing.

Your View
- Look at the big picture.
- Accept change in a changing workplace.

Your Work Commitment
- Always strive for excellence.
- Keep on top of your work.
- Do thorough, careful work.
- Keep developing new skills.

Your Persuasiveness
- Learn how to sell your solutions to problems.
- Spring up after setbacks.
- Demonstrate your creativity.
- Control costs.

know you heard what was said. A lot of enthusiasm goes a long way. Have you ever sat in a meeting where the boss announces some new plans and everyone sits there like a log? Think about how that feels. We all like to receive credit for the good things we do, but often have difficulty accepting blame. Always be accountable for your own actions. Be prepared to accept blame, take responsibility for mistakes, and then correct the problems that caused the mistakes. Share the credit for work well done. While being singled out for praise is satisfying, the best way is to always respond with appreciation for the help of others who were also involved. Work hard at being a team player. Companies cannot have a bunch of cowboys riding off like the "Lone Ranger." Teamwork is essential, and teamwork means real cooperation.

Your accessibility to your boss is important. Your willingness to assist the boss, to do what needs to be done, is always appreciated, even when it's not remarked on. Be available. Getting to work before the boss and leaving after the boss leaves is one of the oldest tips in the business world. Demonstrate your loyalty. Be supportive, never criticize your boss to others, show your willingness to defend and support. Learn good timing. Don't make demands when fires are burning. Be available when there is a likely need.

Your view of what is going on is critical. How you look at things will impact on how your boss deals with you. Be sure to look at the big picture. Even though you know that a Supervisor's work almost always must focus on tactical problems, remember to keep looking at the big picture. In that way, you will demonstrate your understanding of the broader level of management in which your boss is operating. Accept change in a changing workplace. Mergers, acquisitions, spin-offs, and leveraged buy-outs are creating a process of continual organizational change. Technology and changing demographics are having a similar impact. As a manager, your boss is a change agent, and you are the first line person who must be relied on to accept and implement necessary changes.

Your work commitment tells your boss a lot about you. Your boss is more interested in your work commitment than almost any other of these variables. That's because your work commitment is what will get things done, since you are the person dealing with the employees—those who actually make, build, sell, drive or otherwise create the goods and services that companies produce. Always strive for excellence. Perfection is not possible, but excellence certainly is. Keep on top of your work. A huge backlog doesn't necessarily mean you're overworked. It may simply indicate a need to develop your managerial skills. Do thorough, careful work. Thorough, careful work is the only path you can take to reach excellence. Keep developing new skills. Learn a new processing system. Pick up computer skills. Learn planning techniques. The acquisition of new, useful skills must be a never-ending process.

Your persuasiveness is a major part of your potential for success. If you have good ideas, it is important to be able to sell them to your boss. Learn how to sell your solutions to problems. Be persuasive, but be a team player. Spring up after setbacks. A little bounce-up after a fall is always impressive. Demonstrate your creativity. Suggest some innovations, new approaches to problems. Always express your concern for the importance of controlling costs. Most bosses hate managing budgets, but it is what they must do. Those who help them keep costs under control are their natural allies.

Finally, as you progress up the ladder, be sure to groom your successor. Your boss is always going to be concerned about who will replace you. If you have been developing someone, you're one important step ahead.

By now you should have noticed that a lot of these tips for getting along with your boss have a lot to do with your working harder and smarter. That's the real secret.

Disagreeing with Your Boss

Getting along with your boss doesn't always mean agreeing with his/her position. Disagreeing is often a very disagreeable business, but sometimes absolutely necessary. Of course, there is a right way to disagree or provide alternate solutions to workplace problems. We offer you a few rules for your guidance in disagreeing with the boss (Figure 14 – 5).

Rule # 1. Never, never embarrass or put your boss in a corner. Telling your boss in the middle of a group meeting that his/her idea is a bad one probably won't ever work. Tactfully offering an

Figure 14 – 5. The Gentle Art of Disagreeing with the Boss

Rule # 1: Never, never embarrass or put your boss in a corner.

Rule # 2: Get to know when and when not to argue a point.

Rule # 3: Always know what issues or topics aren't open for discussion or dispute.

Rule # 4: Learn how to object without being objectionable.

Rule # 5: Treat your boss as important.

Rule # 6: Know your job before you try much disagreeing.

Rule # 7: Schedule meetings to clarify goals and priorities.

Rule # 8: The bottom line: make your boss look good!

alternate solution may, but is best done in private, or in a hand-written note.

Rule # 2. Get to know when and when not to argue a point. Arguing the point in the group meeting might be OK, but chances are, getting the boss alone will be far more effective. When is also a factor of timing. If your boss is best in the morning, then talk it over in the morning. Grabbing the boss in the hall on the way to an important meeting when he/she is preoccupied and rushed isn't good timing.

Rule # 3. Always know what issues or topics are and aren't open for discussion or dispute. For example, if your boss has strong feelings about all Supervisors wearing suits, it won't be worth it to try to convince her/him otherwise. However, if your boss is unsure about implementing flex time, you may be able to convince him/her that it is a good idea.

Rule # 4. Learn how to object without being objectionable and to disagree without being disagreeable. If you propose an alternate plan to your boss, and it doesn't fly, don't go away pouting. If you convince your boss that your idea is better than hers/his, share the credit and don't gloat.

Rule # 5. Treat your boss as a very important person. Everyone likes to feel important. If you do that, he/she will be prepared to accept some disagreement from you.

Rule # 6. Get to know your job, what you can accomplish and the work environment before you try too much influencing or disagreeing with your boss. If you really know your job, your boss will respect that knowledge and listen.

Rule # 7. Schedule meetings with your boss to establish or clarify goals and priorities—yours and his/hers. This tends to at least reduce disagreement on priorities.

Rule # 8. Remember the bottom line: make your boss look good! It's a lot easier to take advice from a friend.

TIME MANAGEMENT

Career development involves getting things done by effectively using your time. This means using your time to meet priorities in your current position and finding time to do some career planning. Everyone at one time or another has offered this favorite lament: *There just aren't enough hours in the day to get everything done I want to get done.* It is a safe bet that no matter what you do there won't be more than 24 hours in the day. Consequently, you're better off learning how to manage the time you have. Time management is very personal, and what works for you may not work for someone else. The techniques offered here are suggestions. Try them out for a while, then pick and choose what works best for you.

Assess How You Use Your Time

The best way to figure out how you're using your time is to keep a log for a few days or weeks. You don't have to run out and buy some expensive system. Try a pocket calendar, or a simple spiral pocket notebook. Figure 14 – 6 illustrates this method, showing a few frustrating hours in the day of Gillis Jones, Day Supervisor of Sections D1 and D2 of Everybody's Plant.

6:30 breakf. paper. hot news say already 78.

7/13/89 - Thurs.

arr. 8 am plant hot. called engineer to open vents. AC sure would be nice.

8:15 walkthrough D1 and 2. Joes wife to hospit. today. He needs to lv. early, prob. 2:30.

#2022 lathe not working right. Tried to fix it. Took half hour but no go.

Talk to Jim about being late again. This is third time, Shld have given written notice last time. Try to get to it today.

Cary and Joyce upset abt something, but won't tll.

8:40 Damn garbage bn. not picked up last night. Got Jim to help me clean up. He is a good guy, just always late. Don't want to hurt him.

Production stoped. in D2, shipment hasn't arrived. Back to office. It's already 9 am and havn't finished walkthrough

9:15 Can't reach shipper.

9:35 still trying. 15 people out there with nothing to do.

9:45 reached them, they say its on our loading dock doors locked last night.

9:55 clock. yep

10:30 D2 crew helped me move it all in so they can get started now. 8 to 10:30 everyone here and no producth.

Now they'll take their break. Union contrct.

Figure 14 – 6. A Simple Time Log

After you have kept a log for a few days, study it. Ask yourself some questions:

- What did you spend your time doing that you didn't really have to do?
- What could you have delegated instead of doing yourself?
- What did you do to waste other people's time?
- Did you spend your time on high-priority or low-priority items?

Look at Figure 14 – 6 again. What time management mistakes did that Supervisor make in the few hours recorded? Did Gillis spend his time doing what he didn't really have to do? What sorts of things? Could Gillis have delegated instead of doing all those things himself? What, for example? Did Gillis waste other people's time, too? How?

Did Gillis spend his time on high-priority or low-priority items? Do you ever find yourself acting like Gillis?

Time Management: The Plan

If you want to manage your time, you will need to plan. For the purpose of time management, divide the planning into two parts: goals and time utilization.

Identify short- and long-term goals. What would you like to do better with your time? Perhaps your goal is *I really need to be able to handle more of my paperwork.* Maybe it's *I don't want to spend as much time out on the floor.* Or, you feel *I must really increase the time I have to spend with individual workers.* All of these are legitimate goals, and all will take a rearrangement and reallocation of your time. How do you do it?

First, put the goals in priority order. What is most important, second, and so on. Then list the steps that have to be taken to meet these goals. For example: *I have to learn to use the personal computer in order to do my paper work more effectively; I can sign up for a course in August; my boss has already sent in a purchase order for the equipment; I could be underway by September.* Prioritize the steps and work on the most important and achievable first.

Learn better methods of time utilization. Another form of planning is to make a list every week of your activities. Determine the priorities on the list and label them. Determine the least important activities and label them A, B and C with A being the most important and achievable, B the second and C the least. For obvious reasons, this is often called the ABC Time Management Method (Figure 14 – 7). Having identified your priorities, you then spend the majority of your time on the items you have identified as priorities. You do not move from A items until all that can be completed have been completed. After that, you go to B items. C items usually have to wait quite some time, unless something happens to change your priorities.

Time Management: The Tips

Tips for time management usually deal with what are called *time traps.* Time traps are traps that use up your precious time in an

Figure 14 – 7. ABC Time Management Method

Goals	Priority		
	A	B	C
Reduce time spent on paper work	X		
Register for community college course		X Aug	
Learn computer budget systems		X Sept	
Get machineroom equipment serviced	X		
Order parts for binding machine	X		
Purchase replacement binder			X
Increase production on Xeres 503 drills	X		
Order Xeres 503 special booster	X		
Repaint lunchroom		X	

unproductive manner. Gillis Jones caught himself in a time trap when he spent so much time making telephone calls. He had a lot of workers standing around. One of them could have done some calling and Gillis could have been getting more of his work accomplished. To avoid time traps, follow these time-management tips.

- Keep your priorities list within sight.
- Do the most important items first to reduce the stress and distraction and thus time wasted in putting them off.
- Work on top priorities in your peak time. If you're best in the morning, do them in the morning.
- Procrastination is a time trap. If it's a priority item, do it now. If it's something you can do on your way to something else, and will take up no significant time, do it now.
- Don't overschedule your time. Leave time for the unexpected. Do save yourself some lunch and break time. You might even take a little of that time to smell the roses.
- Schedule a meeting with yourself from time to time. Close the door and ask not to be disturbed.
- Clear the unimportant tasks from your desk so you are not distracted.
- Use a calendar or note tickler system (you can get both in most drug, variety or office supply stores).
- Break larger tasks into smaller pieces. It is easier that way and gives you a sense of accomplishment.
- Don't be a perfectionist. We are all human, so none of us is perfect.
- Set target dates for the completion of tasks or projects.
- Don't handle paper twice.
- Delegate, delegate, delegate.
- Take time to plan: one hour of planning saves three to four hours of execution time.
- Try to keep an even pace. Don't burn yourself out by noon.
- Learn to say NO.

As you master these tips you will also begin to manage your time.

WELLNESS

What does wellness have to do with Supervisors and career development? It is difficult to do your job and to develop yourself and your career if you are not at your best, physically and mentally. Not long ago doctors, modern medicine and drugs were responsible for keeping people well. People have learned, however, that they must take responsibility for staying healthy. Doctors simply can't do it for them.

The 10 leading causes of death among Americans all have associated risk factors that the individual can personally control. One's lifestyle can either keep one well or make one sick. For example, cardiovascular disease (the leading cause of death among Americans) has several primary risk factors:

- Smoking
- High blood pressure
- High cholesterol
- Lack of cardiovascular exercise
- Heredity
- Age

Obviously, you can't do anything about your heredity or age, but you can:

- Choose whether or not to smoke
- Learn ways to help control blood pressure
- Maintain a low-fat/high-fiber diet
- Participate in aerobic exercise

All of these measures have a major impact on avoiding heart disease or reducing its impact after you've gotten it.

Controllable Wellness Factors

The following areas are all within your control and can help you be a peak performer.

Nutrition. What you eat is up to you. Maintaining a sound, balanced diet doesn't mean you have to eat only wheatgerm and green leafy vegetables. The four basic food groups are back in fashion and provide a simple framework for what is a balanced diet. The American Heart Association provides good guidelines for choosing within the food groups. You should also have your cholesterol checked to determine if you should pay particular attention to your fat intake. Since one of the major things Americans need to reduce is the volume of food they eat and the fat and cholesterol content of that food, the odds are you'll need the test.

Smoking. If you are a smoker, you know that there is no doubt that smoking is unhealthy. So are other things? That's true. Walking in front of a fast-moving truck comes to mind. Do you smoke to alleviate stress? Did you know that smoking actually contributes to stress? If you smoke, consider trying to quit. It is not only unhealthy for you, but for those around you. With the very rapid change in American attitudes toward smoking, it is already not allowed in many workplaces. Smoking is not simply bad for your health, it is also becoming socially unacceptable. That's quite a change from the Humphrey Bogart movies Hollywood gave us. Contact your local Lung Association, American Cancer Society, or community hospital for smoking cessation programs. If you have tried to quit before and started smoking again, don't be discouraged. Studies show that it usually takes three to four attempts at quitting before you are successful. Most addictions do.

Exercise. Aerobic exercise for 30 minutes, three to four times a week is recommended. Aerobic exercise includes activities such as swimming, bicycling, jogging, rowing and walking. Choose an exercise that is the most enjoyable for you. If you dread exercising, set small goals for yourself and work your way up to the recommended level of exercise. If you overdo it the first few

Figure 14 – 8. Symptoms of Stress

Job-Related:

Frequent errors
Trouble concentrating
Can't make decisions
Dulled motivation

Emotional Symptoms

Unwanted thoughts
Fears and insecurities
Embarrassment or guilt
Negative cynical mood
Low self-respect
Temper outbursts
Loss of joy, depression
Constantly argumentative

Medical Symptoms

Nervous, tense, jumpy
Tired, energy slump
Excesses
Burnout

Figure 14 – 9. Burnout Symptoms

Negative Self-Image

Significant self-doubt
Blaming self
Impatient
Feeling guilty
Sense of failure

Antagonistic and Negative Toward Others

Feeling cynical
General negative attitude
Feeling bitter and sarcastic
Feeling defensive

Alienated and Unconnected

Loss of sense of purpose
Blaming others
Feeling immobilized
Being bored or overly rigid

Withdrawn and Resigned

Feeling depressed
Avoiding others
Feeling listless
Being absent in the present

times, you may be too discouraged and sore to continue. Whatever you do, if you haven't been exercising, contact your physician before starting any type of exercise program.

Driving. Motor-vehicle accidents are the leading cause of death in Americans aged 5 to 34 years and the seventh leading cause of death overall. The Department of Transportation postulates that universal use of safety belts would reduce motor-vehicle-accident-related fatalities by 50% and injuries by 65%. (*Journal of the American Medical Association,* December 23/30, 1988.) So, buckle-up! Given the statistics, driving under the influence of alcohol is a death wish. The results of drinking and driving have been hammered home in television shows and advertising. Still, the death toll goes on. Don't contribute to it.

Stress Management

An effective wellness program must deal with stress management. Supervisors experience stress just like their employees, maybe worse. The factors of employee stress discussed in Chapter 13 affect Supervisors as well. In addition, Supervisors are particularly prone to these additional types of stressors: work overload, role conflict, job ambiguity and dual responsibilities.

Recognizing stress symptoms. If you are going to control stress in yourself, you will need to watch for its symptoms. Figure 14 – 8 lists common stress symptoms. They may be work behaviors, emotional behaviors or medical symptoms. Part of staying well is watching for symptoms and then taking appropriate action. That is precisely what you must do with stress. Study the chart well.

Burnout. Burnout is the worst-case scenario for stress in the workplace. Burnout generally means someone who is exhausted, used up or worn out by excessive use. The underlying causes of burnout are shown in Figure 14 – 9. It's not a pretty picture, and using stress management and stress relief techniques can usually prevent you from going that far.

Stress relief techniques. Following the above wellness tips will help you manage stress: maintaining good nutrition, not smoking, exercising regularly and driving safely (as well as wearing your seat belt and never drinking and driving). There are also a good many relaxation techniques such as deep breathing, head rolls, stretching, etc. for reducing or relieving some of the immediate physical and emotional symptoms of stress. Everyone has a favorite way to reduce stress—quiet music, fishing, cooking, camping, boating, reading, walking, and so on. But it's always helpful to learn a few more. Figure 14 – 10 is a checklist for stress reduction developed by Barbara Hanson Dennis for a stress reduction program for Supervisors and managers. You may find many of the tips quite helpful.

Counseling. Take advantage of counseling from your Employee Assistance Program or stress management classes offered by your company or available to you from other organizations. To learn more about taking charge of your lifestyle, take advantage of these resources:

- Your company's Wellness Program or EAP

Figure 14 – 10. Daily Checklist for Control and Stress Reduction

1. Work to improve what you do best and most readily.

2. As much as possible, rely on yourself to accomplish the goals you have set for yourself.

3. Concentrate first on your highest priorities.

4. Don't succumb to the feeling that you don't have time to do what you want.

5. Prepare a general schedule the night before, but approach each day in a relaxed way, letting things emerge and evolve as the day goes on.

6. When you finish one activity, move on to another.

7. Seek activities that you enjoy.

8. Focused and informed activity reduces fear and anxiety. Studying and testing a task will take the sting out of failure.

9. Criticism, however unpleasant, can provide valuable information about ways to improve.

10. Don't dwell on potential difficulties that are beyond your control.

11. Postponement can become habitual and lead to nonproductivity.

12. Don't allow yourself to be distracted by opportunities for self-indulgence.

13. Don't resort to mechanical formulas to solve problems. New situations require new solutions.

14. Don't blame your inaction on others or take credit for sacrificing your goals on their behalf. This demeans them and creates insecurity about your true feelings.

15. Act in terms of your personal goals to minimize the psychological threats of specific situations.

16. Stick to what you find most rewarding.

17. Remember that the strongest relationships develop from pursuit of a common objective or activity.

- Community hospitals or medical centers
- Nonprofit associations such as the American Cancer Society, American Lung Association and the American Heart Association
- Your personal physician

CASE STUDY: MOVING UP

Two first-line Supervisors had applied for a middle-management position with the Gray Steel Company. Both of the Supervisors had similar years of experience. In the interview, however, their stories were a little different. Gray Steel was looking for someone who would not only excel in the current position but also showed potential for future advancement. Here is part of both interviews. Whom would you choose?

Interview 1:

Interviewer: *Samantha, it looks like you've had a lot of*

good experience as a Supervisor. Why are you interested in this position?

Applicant: *I've spent a lot of time thinking about my career and my strengths and weaknesses. I've realized that I am a good Supervisor and that I can apply these skills to a management position. I've taken some training courses in planning, budgeting and leadership, and I really enjoy these management responsibilities. I'm always looking for a new challenge, and I think that's what this position will provide.*

Interview 2:

Interviewer: *Melissa, it looks like you've had a lot of good experience as a Supervisor. Why are you interested in this position?*

Applicant: *Well, I've been a Supervisor for a few years now, I like the job and the company, and I figure it's time to move on. This position just seems like a natural next step. Besides, I could really use the extra money.*

How would you respond in this type of situation? Are you making plans and taking the appropriate steps for your next career move? Or, are you waiting for the next opportunity to happen along?

CONCLUSION

This chapter explored the development of your career as a Supervisor, how you can enhance that job, and how you can seek promotions and new careers. To do that, you need to take control of your career by assessing your priorities and goals, designing a development plan/strategy, learning how to influence and deal with your boss, and practicing time management and personal wellness. These approaches should open up new career opportunities both as a Supervisor and in other aspects of your worklife.

Bibliography

The following books are recommended by the authors for further reading in the areas indicated.

CAREER DEVELOPMENT

Hagberg, Janet. *The Inventurers: Excursions in Life and Career Renewal.* Reading, MA: Addison-Wesley Publishing, 1988.

Schein, Edgar H. *Career Dynamics: Matching Individual and Organizational Needs.* Reading, MA: Addison-Wesley Publishing, 1978.

COMMUNICATION

Cherney, Marcia B. et al. *Communicoding.* New York: Donald I. Fine, 1989.

DISCRIMINATION AND EMPLOYMENT LAW

Morrison, Ann M. and Mary Ann VonGlinow. "Women and Minorities in Management," *American Psychologist,* February, 1990.

Hunt, James W. *The Law of the Workplace: Rights of Employers and Employees.* 2nd ed. Washington, D.C.: Bureau of National Affairs, 1988.

Kalet, Joseph E. *Age Discrimination in Employment Law.* Washington, D.C.: Bureau of National Affairs, 1986. (Frequently updated.)

Twomey, David P. *A Concise Guide to Employment Law, EEO and OSHA.* Cincinnati, OH: South-Western Publishing, 1985. (Frequently updated.)

LABOR RELATIONS AND LABOR LAW

Fossum, John A. *Labor Relations, Development, Structure, Process.* 4th ed. Homewood, IL: BPI/Irwin, 1988.

McGiness, Kenneth C. and Jeffrey A. Norris. *How to Take a Case Before the NLRB.* 5th ed. Washington, D.C.: Bureau of National Affairs, 1986.

MULTICULTURAL WORKFORCE

Adler, Nancy J. *International Dimensions of Organizational Behavior.* Boston, MA: Kent Publishing Company, 1986.

Copeland, Lennie, and Lewis Griggs. *Going International: How to Make Friends and Deal Effectively in the Global Marketplace.* New York: Randon House, 1985.

Harris, Philip R., and Robert T. Moran. *Managing Cultural Differences: High-Performance Strategies for Today's Global Manager.* 2nd ed. Houston, TX: Gulf Publishing, 1987.

Schaupp, Dietrich L. *A Cross-Cultural Study of a Multinational Company: Attitudinal Responses to Participative Management.* New York: Praeger Publishers, 1978.

PERFORMANCE

Fournies, Ferdinand F. *Coaching for Improved Work Performance.* New York: Van Nostrand Reinhold, 1978.

_____. *Why Emplolyees Don't Do What They're Supposed to Do and What to Do about It.* Blue Ridge Summit, PA: Liberty House, 1989.

SUPERVISION

Bittel, Lester R., and John Newstrom. *What Every Supervisor Should Know.* 6th ed. New York: McGraw-Hill Book Co., 1989.

DuBrin, Andrew J. *The Practice of Supervision: Achieving Results Through People.* 2nd ed. Chicago, IL: Irwin, 1987.

Blanchard, Kenneth P., Ph.D. and Spencer, Johnson, M.D. *One Minute Manager.* Nightingale-Conant, 1989.

TIME MANAGEMENT

Oncken, William Jr. *Managing Management Time: Who's Got the Monkey?* Englewood Cliffs, NJ: Prentice-Hall, Inc., 1984.

TRAINING

Eitington, Julius E. *The Winning Trainer.* 2nd ed. Houston, TX: Gulf Publishing, 1988.

Knowles, Malcolm. *The Adult Learner: A Neglected Species.* 3rd ed. Houston, TX: Gulf Publishing, 1984.

Odiorne, George S. and Geary A. Rummler. *Training and Development: A Guide for Professionals.* Chicago, IL: Commerce Clearing House, 1988.

Phillips, Jack J. *Handbook of Training Evaluation and Measurement Methods.* Houston, TX: Gulf Publishing, 1983.

WAGES AND HOURS

Kalet, Joseph E. *Primer on Wage and Hour Laws.* Washington, D.C.: Bureau of National Affairs, 1987. (Frequently updated.)

WELLNESS

Blanchard, Marjorie, Ph.D., and Mark J. Tager, M.D. *Working Well: Managing for Health and High Performance.* New York: Simon and Schuster, 1985.

WRITING

Strunk, William Jr., and E. B. White. *The Elements of Style.* 3rd ed. Macmillan, 1979.

Index